SRI YANTRA

step by step

Draw in AutoCAD

any 2D drafting engineering application

Ashwini Kumar Aggarwal

जय गुरुदेव

ISBN13: 978-93-95766-62-3 Paperback Edition
ISBN13: 978-93-95766-58-6 Hardbound Edition
ISBN13: 978-93-95766-61-6 Digital Edition

FULL COLOR PLATES

Title: **Sri Yantra step by step draw in AutoCAD**
Author: **Ashwini Kumar Aggarwal**

Printed and Published by
Devotees of Sri Sri Ravi Shankar Ashram
34 Sunny Enclave, Devigarh Road,
Patiala 147001, Punjab, India

https://advaita56.in/
The Art of Living Centre

https://www.artofliving.org/

30th June 2023, DevShayani Ekadashi Parana, Beginning of Chaturmas, Sarvartha Siddhi Yoga, Ashadha Masa, Shukla Paksha, Dvadashi Tithi, Vishakha upari Anuradha Nakshatra, Grishma Ritu, Uttarayana. 181st day of this year in Gregorian calendar.
International Asteroid Day. Meteor Watch Day.

On this day in 1893 Bluish white pear-shaped Excelsior Diamond discovered in Jagersfontein mine in South Africa.
1908 Tunguska Impact in Siberia when a large meteoroid entered Earth's atmosphere and then detonated in the sky.
1936 Romantic novel - Gone with the Wind by Margaret Mitchell published.

Vikram Samvat 2080 Pingala, Saka Era 1945 Shobhakrit

1st Edition June 2023

जय गुरुदेव

Dedication

Sri Sri Ravi Shankar
who gave us **Devi Puja sacred geometry** in the form of **Navratri Chandi Homa structures**

Acknowledgements

Superb visit to Shimla, long treks in the nourishing purifying healing climate and pure Oxygen.

We acknowledge the pioneering efforts of Patrick Flanagan and Paul Delisle who threw light on drawing a Sri Yantra.

Disclaimer

AutoCAD is a registered trademark of Autodesk Inc. Used here for identification purposes only and does not imply endorsement.

Blessing

All the Saints in the past, when they went deep into meditation, they just heard Om. So, Om means many things. It means love, eternity, purity, peace. Om is made up of several dhatus: 'Ah' 'Oo' 'Ma'. Just 'Ah' has 19 meanings.
The total prana is represented by one syllable OM. Before birth, we were part of that sound and after death we will merge with that sound; the SOUND of the Spirit.

Sri Sri Ravi Shankar
Birthday Celebrations, Montreal Q & A in Canadian Ashram, May 10, 2012

Prayer

ॐ भद्रं कर्णेभिः श्रृणुयाम देवाः । भद्रं पश्ये माक्षभिर् यजत्राः । स्थिरैरङ्गैस्तुष्टुवाꣳसस्तनूभिः । व्यशेम देवहितं यदायुः ॥ स्वस्ति न इन्द्रो वृद्धश्रवाः । स्वस्ति नः पूषा विश्ववेदाः । स्वस्ति नस्ताक्ष्र्यो अरिष्टनेमिः । स्वस्ति नो बृहस्पतिर्दधातु ॥ ॐ शान्तिः शान्तिः शान्तिः ॥

Shanti Mantra of Atharvaveda

O Divine Wisdom! May our ears listen to the sacred and the auspicious. May our eyes see the propitious as we come together to partake of wisdom. May our limbs be firm and body attuned to long endurances. May our senses function with full alertness and May the sense of contentment be strong. May our good thoughts form a discus to shield us, and May our education give us a shining personality. Peace in our heart, in our body and in our environs.

Sri Yantra

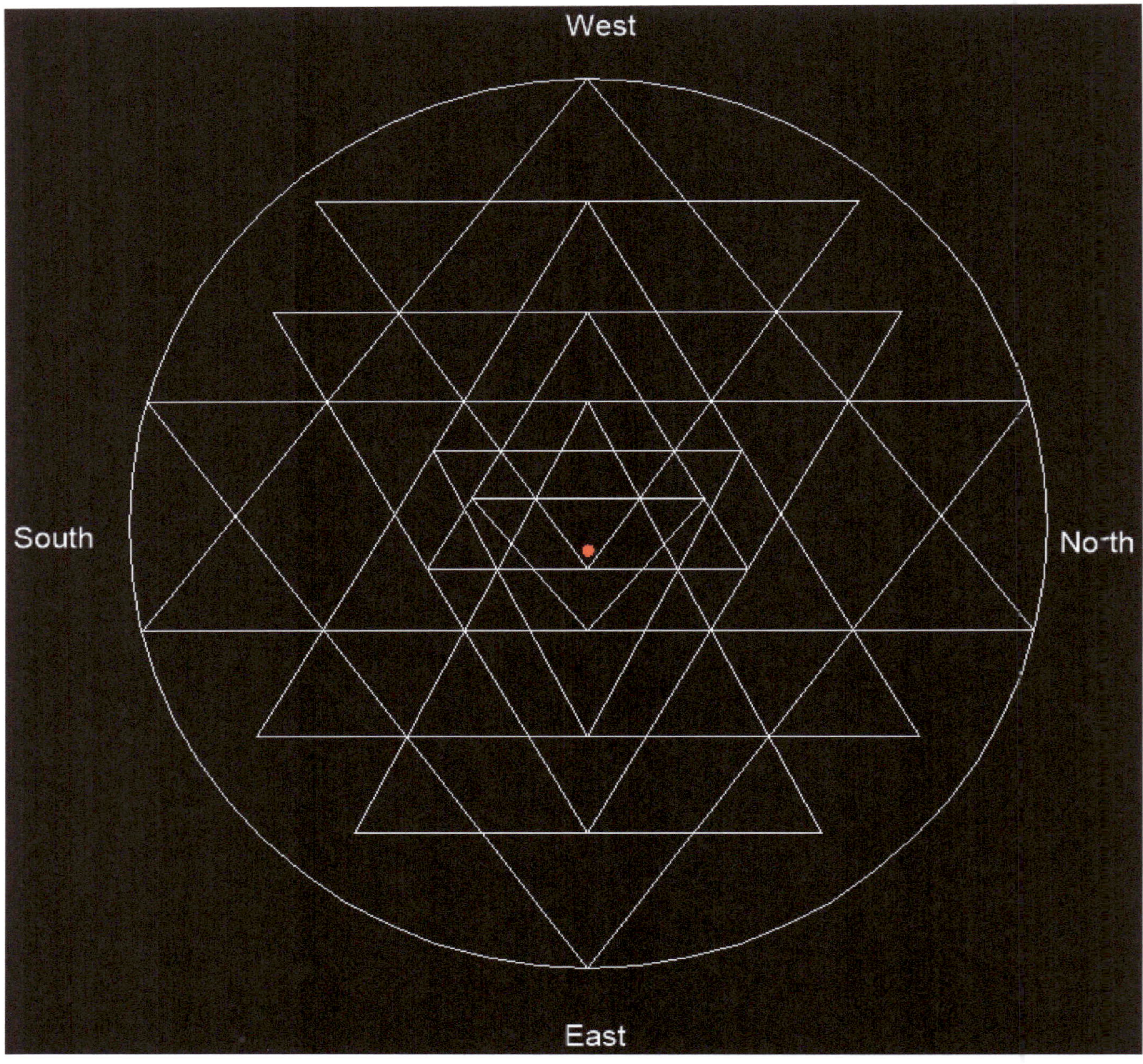

Contents

DIMENSIONS OF SQUARE RECTANGLE 3CIRCLES GOLDEN RATIO TRIANGLE 85

SRI YANTRA ACCURACY AND PRECISION 87

TRIANGLE NAMES FOR CLARITY 94

FOUR UPWARD APEX SHIVA TRIANGLES 103

FOUR DOWNWARD APEX SHAKTI TRIANGLES 104

ONE DOWNWARD APEX UNION TRIANGLE WITH BINDU 105

SRI YANTRA WITH SANSKRIT VERSES AND SEED SOUNDS 106

SANDHI POINTS 24 DUAL INTERSECTIONS 108

MARMA POINTS 18 TRIPLE INTERSECTIONS 109

PROBABLE ERRORS IN OUR SRI YANTRA CONSTRUCTION 110

GOLDEN RATIO TRIANGLE IN SRI YANTRA 112

SRI YANTRA IS FROM BHAVANA UPANISHAD भावनोपनिषद् 114

LEGEND 116

REFERENCES 123

EPILOGUE 124

Table of Figures

Sri Yantra

Preface

What is a Sri Yantra? A healing mandala diagram made by 4 Upward Apex Triangles and 5 Downward Apex Triangles, properly intertwined to represent
- the cosmic forces in creation, and
- the principles of duality, harmony, and divinity.

The 9 Basic Triangles create a total of 43 Triangles due to the overlapping, which are grouped (shaded) as $14 + 10 + 10 + 8 + 1 = 43$.

Apart from these Triangles,
- there is a central Dot (Bindu)
- there are three Circles encircling the Triangles
- there are four Gates encircling the entire Mandala, composed of three lines.

Sanskrit Seed Sounds (bija mantra) like ॐ Om, are potent healing and nourishing energies. Sound is an energy. Pleasing sounds (meditative music, blessings) have excellent effects on one's behavior, attitude, and long-term all round success in life. Harsh and deterring sounds (cruel speech, improper audio-video) wipe away our enthusiasm, drain our energy, block talent and result in suffocation and clouding of neural activity, thereby reducing success.

An aesthetically made Mandala can become vastly superior when infused with Seed Sounds. It is like getting a good vehicle and then employing a proper driver. The Sri Yantra is one of the most ancient mystical, benevolent and highly rewarding mathematical forms of sacred geometry. Recent archaeological and scientific explorations have again and again highlighted that famous and affluent cities of the world e.g., Washington DC, Paris, Bombay, etc. have been built using principles of sacred geometry. Architects and Engineers like Leonardo Da Vinci, Mayan Vishwakarma, etc.; and Structures including the Great Pyramid of Giza, the Taj Mahal, the Shiva Shakti temples of Thanjavur, etc.; have made extensive use of sacred geometry.

This book contains the procedure to draw a Sri Yantra using an engineering software like AutoCAD. It is hoped that modern architects, builders and Yantra manufacturers can employ precise techniques to get it right.

<u>For ease in understanding, we give arbitrary names to the triangles</u>
T1u = Upward Apex Largest Triangle, having Golden Ratio.
T2u, T3u, T4u are the other Upward Apex Triangles in **sequence from top**.
T1d, T2d, T3d, T4d are the four Downward Apex Triangles in **sequence from below**.
T5d downward apex triangle is the innermost triangle enclosing the Bindu.

<u>Also to indicate there are a total of 9 basic triangles we say</u>
1st T1u, 2nd T1d, 3rd T2u, 4th T2d, 5th T3u, 6th T3d, 7th T4u, 8th T4d, 9th T5d.

As each new side gets drawn, we get the sandhi junction points for seeing the Apex or Sides for the next Triangle.

॥ ॐ श्री ललिता महात्रिपुरसुन्दर्यै नमः ॥

श्री यन्त्रम्

Magic of Numbers vis-a-vis the Triangle

Various scriptures describe the **human mind body** complex using different numbers.

The Bhagavad Gita uses the numbers 10, 8 in different verses.
Here in the Sri Yantra, we see there are 10 Outer Triangles of the fifth Enclosure.
Also, there are 10 Inner Triangles of the sixth Enclosure.
And, there are 8 Triangles of the seventh Enclosure.

The Vijnana Bhairava by Swami Lakshman Joo in its commentary of the 54th verse points to 36 elements in creation. Guruji also said this during Maha Shivaratri 2016 at Bangalore Ashram.
If we consider the different aspects of the Sri Yantra, we can discriminate as:
1st Enclosure = 4 Outer Gates = represents the cardinal directions
2nd Enclosure = 16 Outer Petals = represents all vowels
3rd Enclosure = 8 Inner Petals = represents all consonants
7th Enclosure = 8 Triangles = represents the entire Sanskrit Alphabet, all sounds in speech
Sum = 4+16+8+8 = 36 represents all 36 elements (principle or *tattva*) in creation.

The Mandukya Upanishad says that 19 facets can be culled on the physical plane:
i. Bones, ii. Muscles, iii. Nerves, iv. Organs, v. Fluids, vi. five breaths, xi. five senses, xvi. Mind, xvii. Intellect, xviii. Ego, xix. Citta (memory), xx. SELF = 20 aspects in living creation.
Here, 20 – 1 (the SELF) = 19 physical assets.

A Centered Triangular Number given by $(3n^2+3n+2)/2$ results in 19 for n = 3.

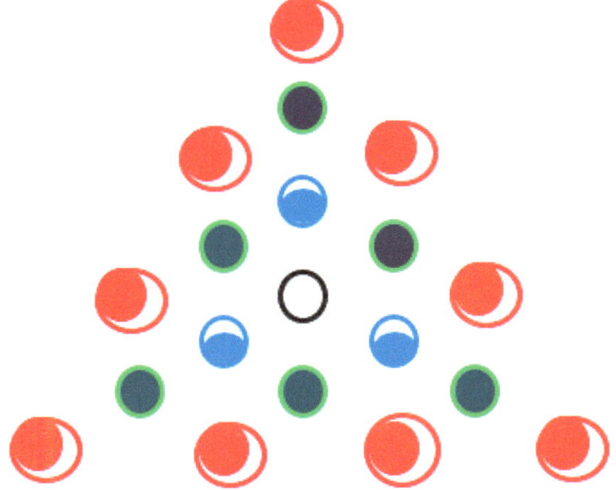

The Sri Yantra consists of many isosceles Triangles.

9th Enclosure Navamā Āvaraṇa - Bindu

निरुपाधिका संविदेव कामेश्वरः ॥ २६ ॥ सदानन्दपूर्णा स्वात्मैव परदेवता ललिता ॥ २७ ॥ लौहित्यमेतस्य सर्वस्य विमर्शः ॥ २८ ॥

nirupādhikā saṃvideva kāmeśvaraḥ ‖ 26 ‖ sadānandapūrṇā svātmaiva paradevatā lalitā ‖ 27 ‖

lauhityametasya sarvasya vimarśaḥ ‖ 28 ‖

That which cannot be contained in form, that which is one's innermost self, is called Kameshvara. And continually filled with bliss, verily one's own innermost self , that which cannot be fathomed is called Lalita. During deep meditation, the ruddy rosy hue of the Self is seen.

श्री सर्वानन्दमय चक्रं नवमावरणम्

Ninth āvaraṇa navamāvaraṇa नवमावरणम् - sarvānandamayacakra श्री सर्व आनन्द-मय चक्रं

Bindu bindusthāna, red color श्रीं

Śrī Mahātripurasundarī (Śiva-śakty-aikya-rūpiṇī)

श्रीललिता महा त्रिपुर सुंदर्यै नमः

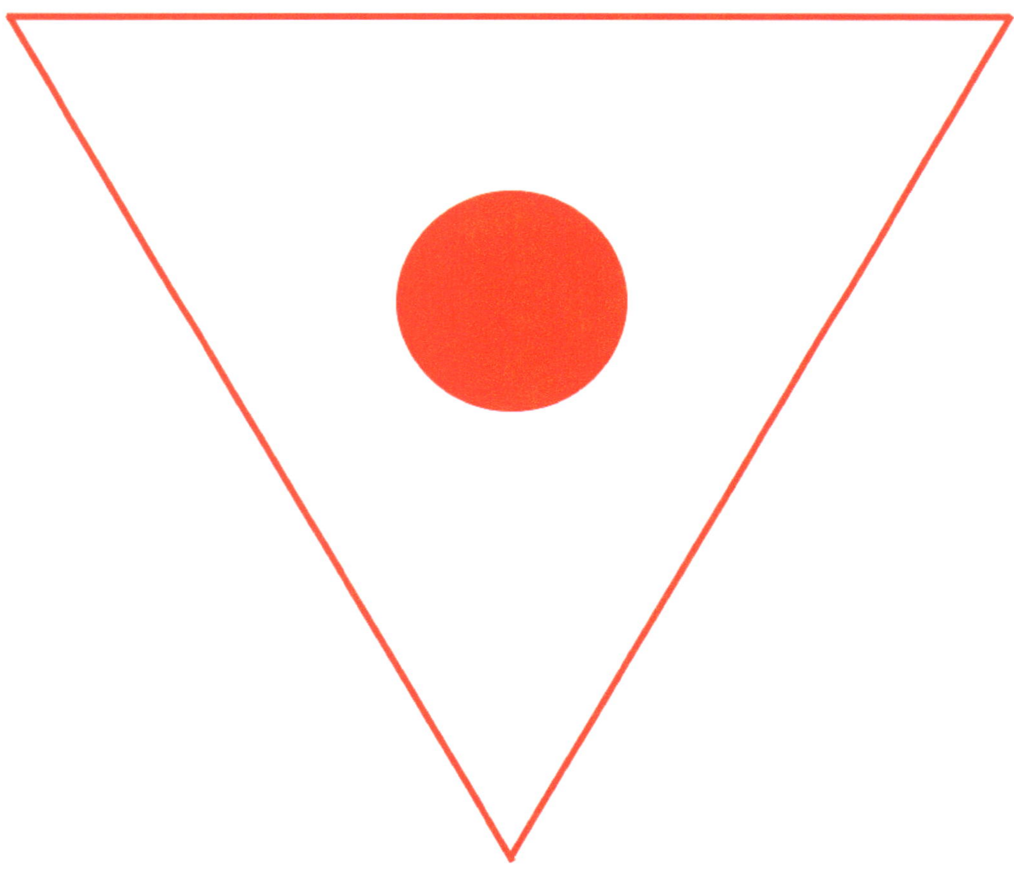

Figure 1 **Sri Lalita Maha Tripura Sundari**

Become Still, Silent, Soft, Melted, Absorbed in Divine Union.

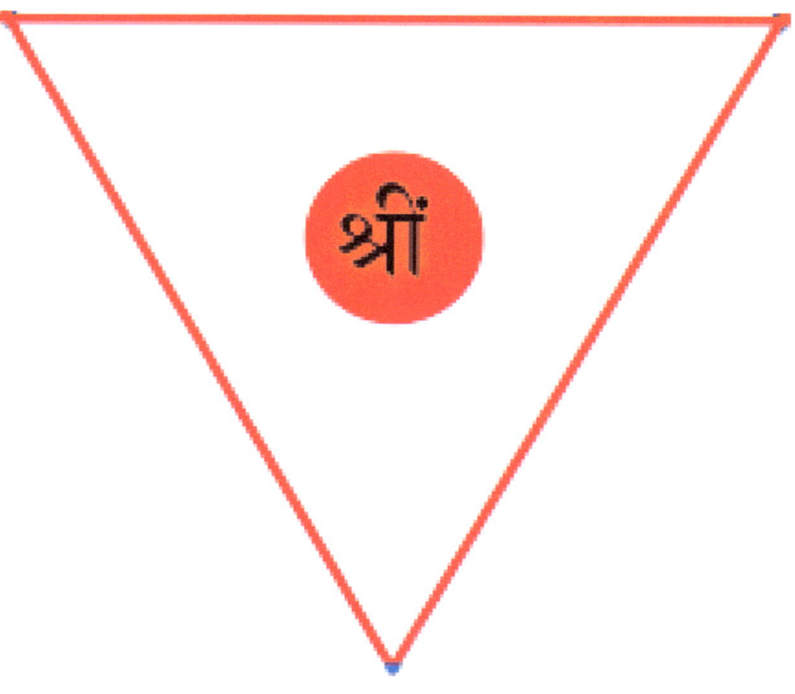

Sri Yantra step by step Draw Guidelines

A Sri Yantra designed with accuracy and precision is precious, beautiful and highly benevolent. To begin drawing, use a 2D drafting engineering software e.g. AutoCAD, BricsCAD, nanoCAD, LibreCAD etc.

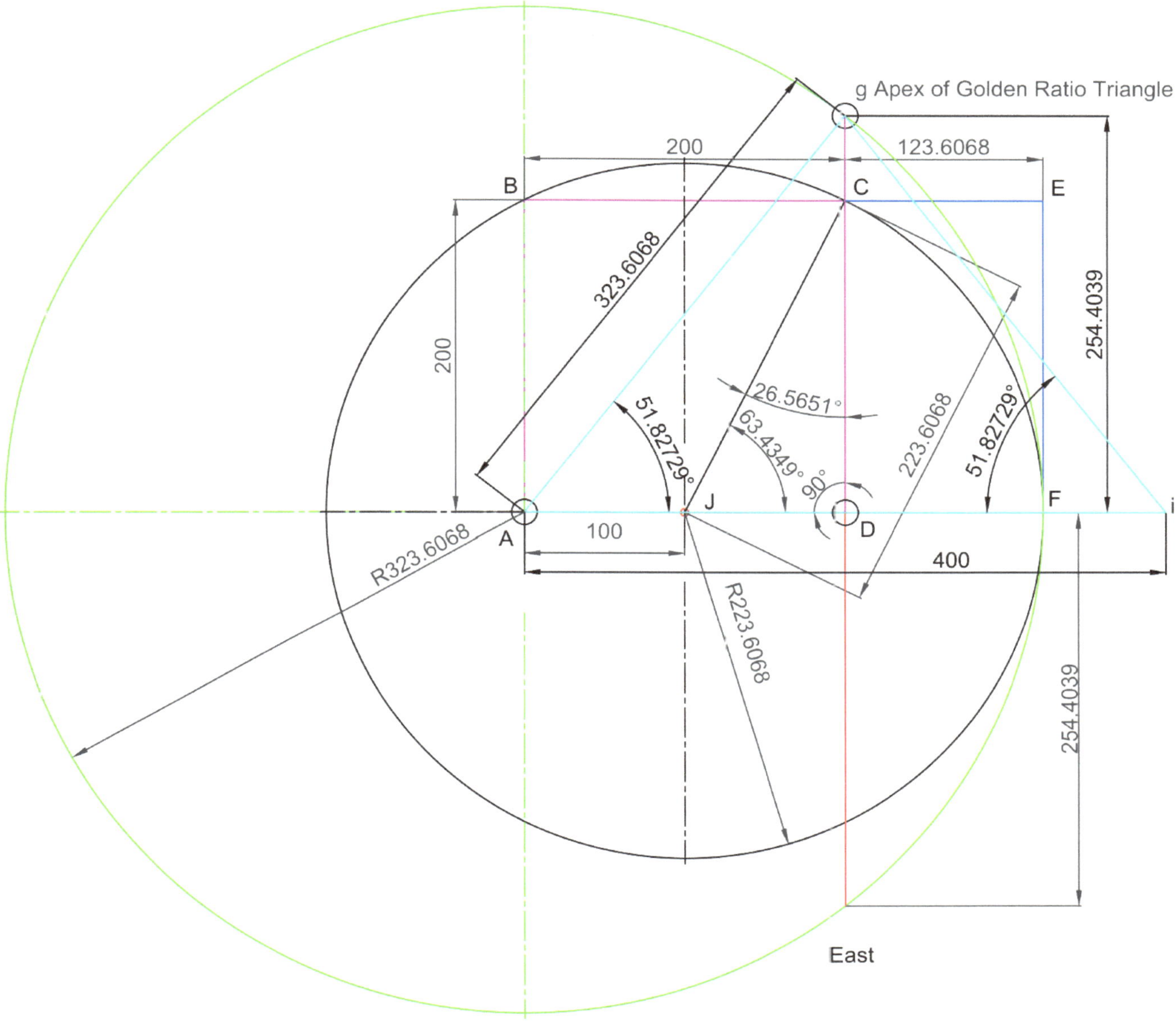

- Set the Dimensions precision to **four decimal** places, e.g., **0.0000**
- With each Line, Angle, or Circle drawn, **take** the Dimension and see if it matches exactly.
- Even a **slight** error shall make the Sri Yantra lose its Beauty and Preciousness.
- 1 Square ABCD, side 200.
- 2 Bisect Square at J,
    - make right angle Triangle JCD with hypotenuse JC = 223.6068, Angle CJD = 63.4349°
- 3 Draw Black Circle with **center at J**, radius = hypotenuse JC = 223.6068
- 4 Extend Square ABCD to a Rectangle ABEF, where longer side AF = BE = 323.6068
- 5 Draw Green Circle with **center at A**, radius = 323.6068
- 6 **Extend** side of Square DC to touch Green Circle at upper point **g**. The length Dg = 254.4039
    - For exactness,

- o   Extend side of Square CD to touch Green Circle at lower point. This length = 254.4039
  - o   These both lengths must be **Equal**. *This is just to check if we are going right so far.*
  - o   If both these lengths are **Equal**, we have found the **Apex g** for our **Triangle T1u**.
- 7 Now **join** A to g, and we get the Side **Ag** of our **Golden Ratio Triangle** T1u.
  - o   Note that **Side** Ag = radius of Green Circle = **323.6068**
- 8 Copy and **Mirror** the Side Ag to get the **Side gi** of Golden Ratio Triangle T1u.
- 9 Join points A to i to get **Base Ai** of Golden Ratio Triangle T1u
- **Golden Ratio Triangle T1u** Largest Upwards Apex Triangle is **starting** Point for **Sri Yantra**.
- **Dimensions** for **Golden Ratio Triangle** T1u = **Agi** must be **precisely**
  - o   Baseline length = 400
  - o   Isosceles Side Length each = 323.6068
  - o   Isosceles Angle each = 51.82729°

1 Begin by Drawing a Line

Figure 2 **Start by drawing a Line**

1. **First** draw a **Line** = AJ = 100

2 Extend to a Square

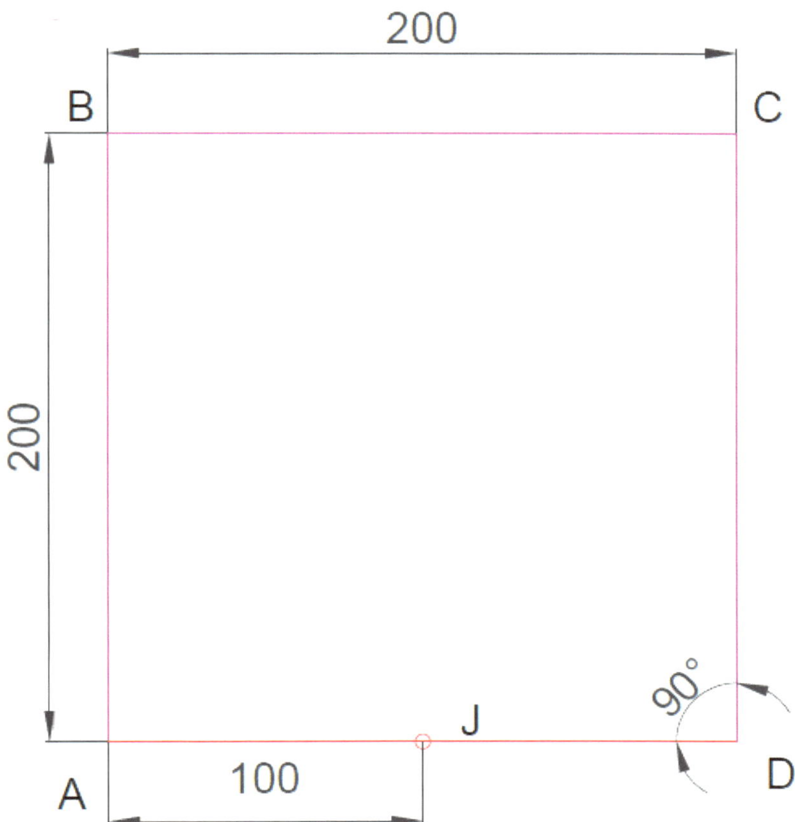

Figure 3 **Draw a Square**

1. **First** draw a **Line** = AJ = 100
2. **Extend** it to draw a **Square** = ABCD, side AB = AD = 100, Point J bisects side AD.

3 Mark Hypotenuse of a Right Angle Triangle

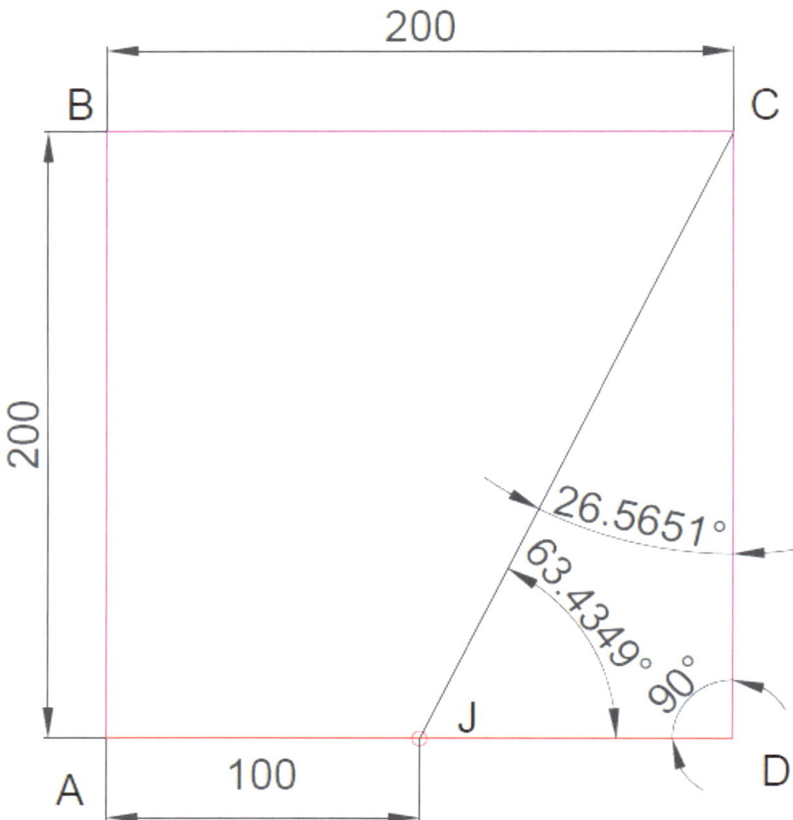

Figure 4 **Draw hypotenuse to make Right Angle Triangle**

1. **First** draw a **Line** = AJ = 100
2. **Extend** it to draw a **Square** = ABCD, side AB = AD = 100, Point J bisects side AD.
3. Join JC to make the **hypotenuse** of a Right Angle Triangle JCD. Angle CJD = 63.4349°

4 Use Hypotenuse to Draw 1st Circle

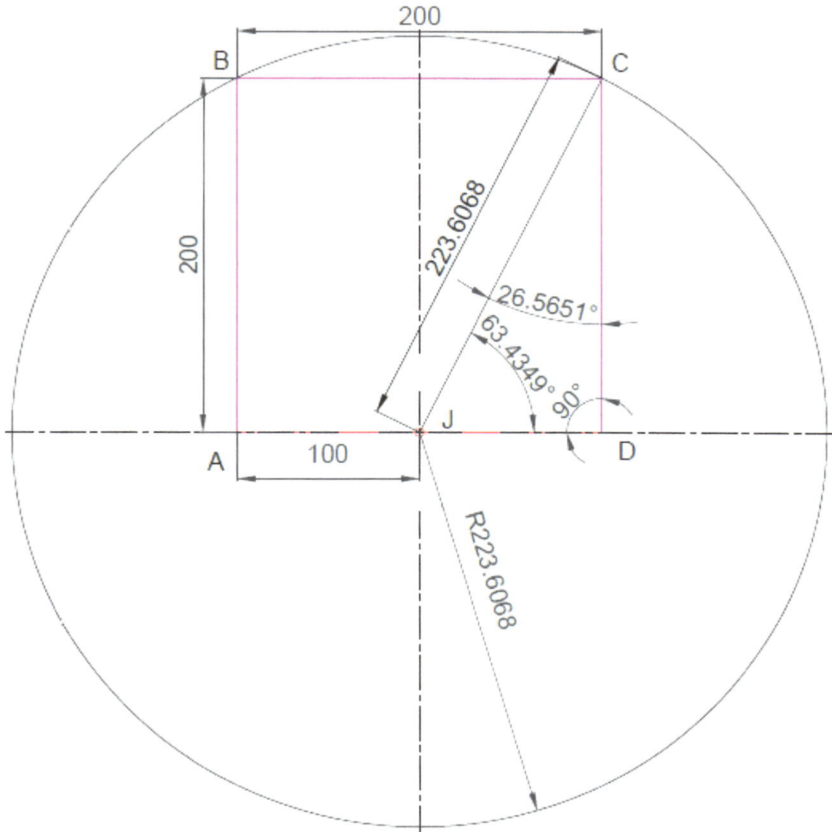

Figure 5 **Use hypotenuse to draw 1st Circle**

1. **First** draw a **Line** = AJ = 100
2. **Extend** it to draw a **Square** = ABCD, side AB = AD = 100, Point J bisects side AD.
3. Join JC to make the **hypotenuse** of a Right Angle Triangle JCD. Angle CJD = 63.4349°
4. Draw **1st Circle** with center J, radius = hypotenuse JC = 223.6068

5 Extend Square to a Golden Ratio Rectangle

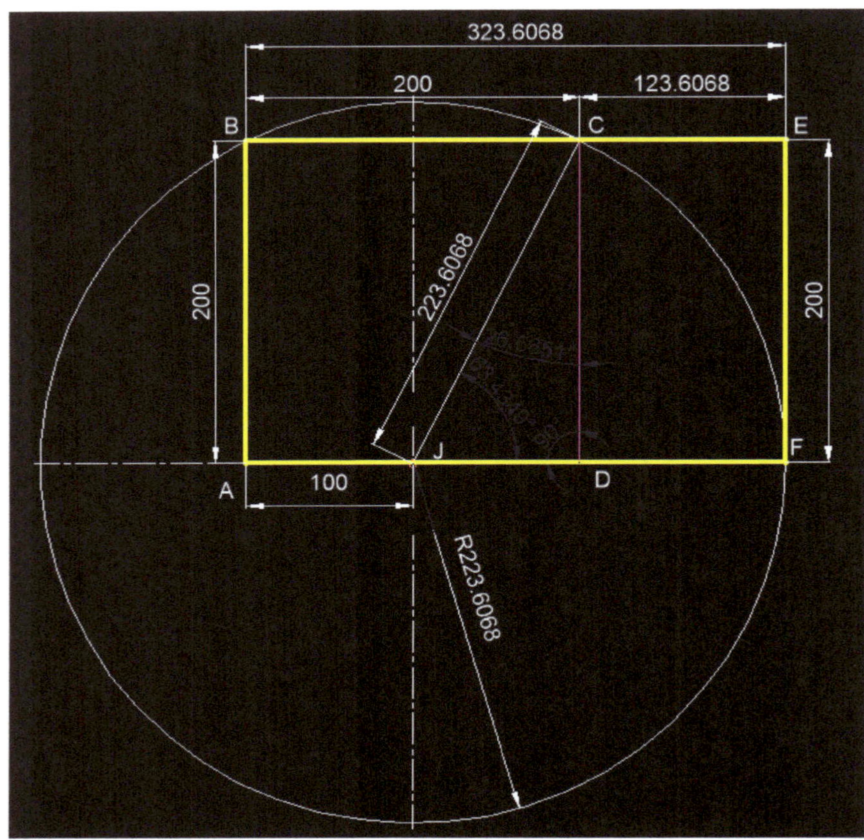

Figure 6 **Golden Ratio Rectangle**

1. **First** draw a **Line** = AJ = 100
2. **Extend** it to draw a **Square** = ABCD, side AB = AD = 100, Point J bisects side AD.
3. Join JC to make the **hypotenuse** of a Right Angle Triangle JCD. Angle CJD = 63.4349°
4. Draw **1st Circle** with center J, radius = hypotenuse JC = 223.6068
5a. Extend AD to touch Circle at Point F. Use segment DF to extend BC to BE, so that a
   Golden Ratio Rectangle is formed.
5b. Dimensions of Golden Ratio Rectangle, longer Side AF = BE = 323.6068, shorter side AB = EF = 200.
   Ratio of longer side/shorter side = 323.6068/200 = **1.618034** = Φ Phi = Golden Ratio.

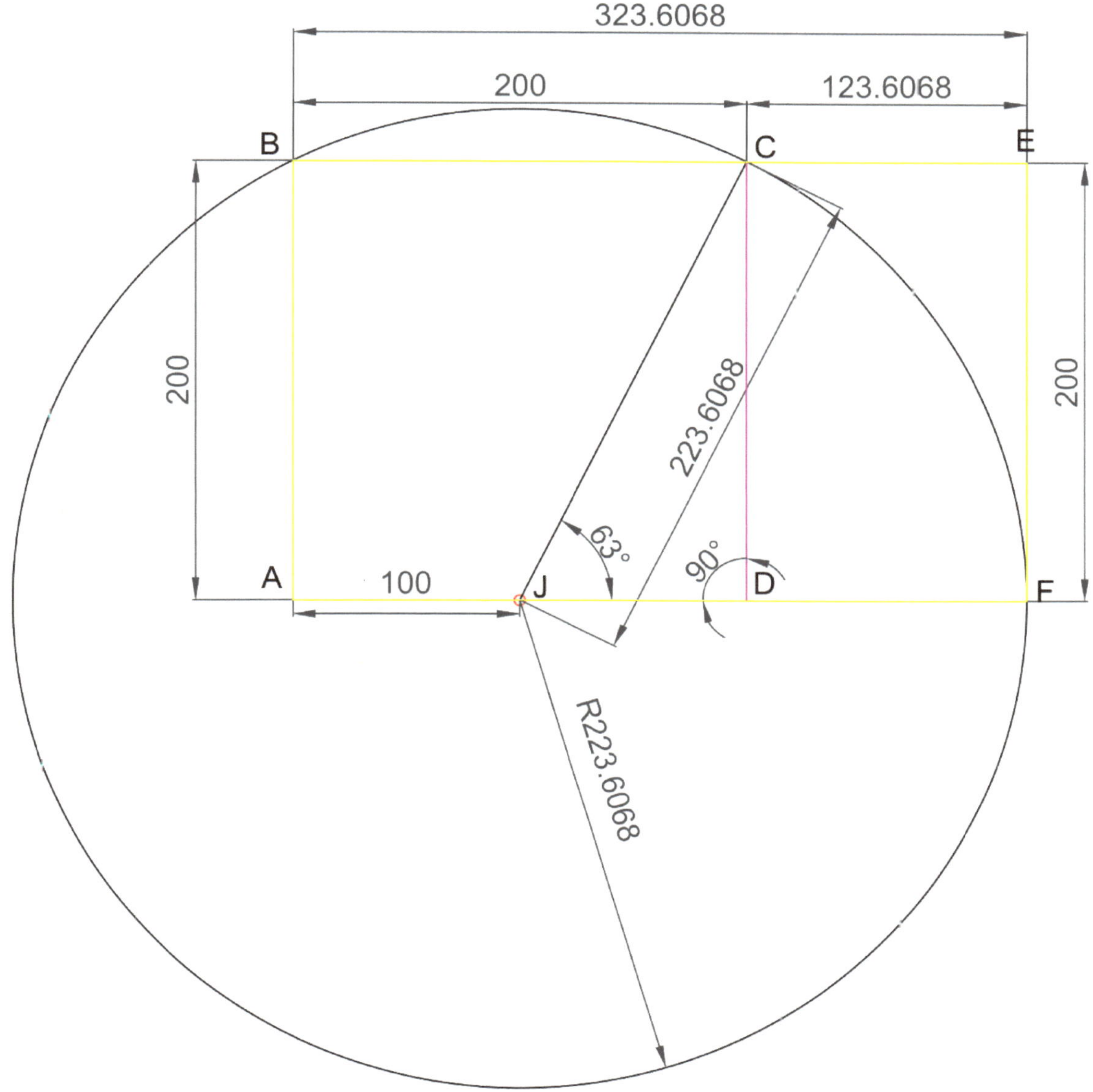

Figure 7 **Golden Ratio Rectangle (another view)**

1. **First** draw a **Square** = ABCD, **side AB = AD = 200**
2. **Bisect** the side AD of Square to get Point **J**
3. **Right Angle** Triangle = JDC, Angle CJD = 63.4349°, base length **JD = 100**, Hypotenuse = 223.6068
4. Draw a **Circle** with **center J**, **radius** = hypotenuse = **223.6068**
5. **Extend** side AD to **touch** Circle at **Point F**.
6a. Draw Golden Ratio **Rectangle** = ABEF, **longer side** AF = BE = **323.6068**, shorter side AB = EF = 200
6b. **Golden Ratio Rectangle** longer side/shorter side = 323.6068/200 = **1.618034** = Φ Phi

6 Draw 2nd Circle to get Apex of Golden Ratio Triangle

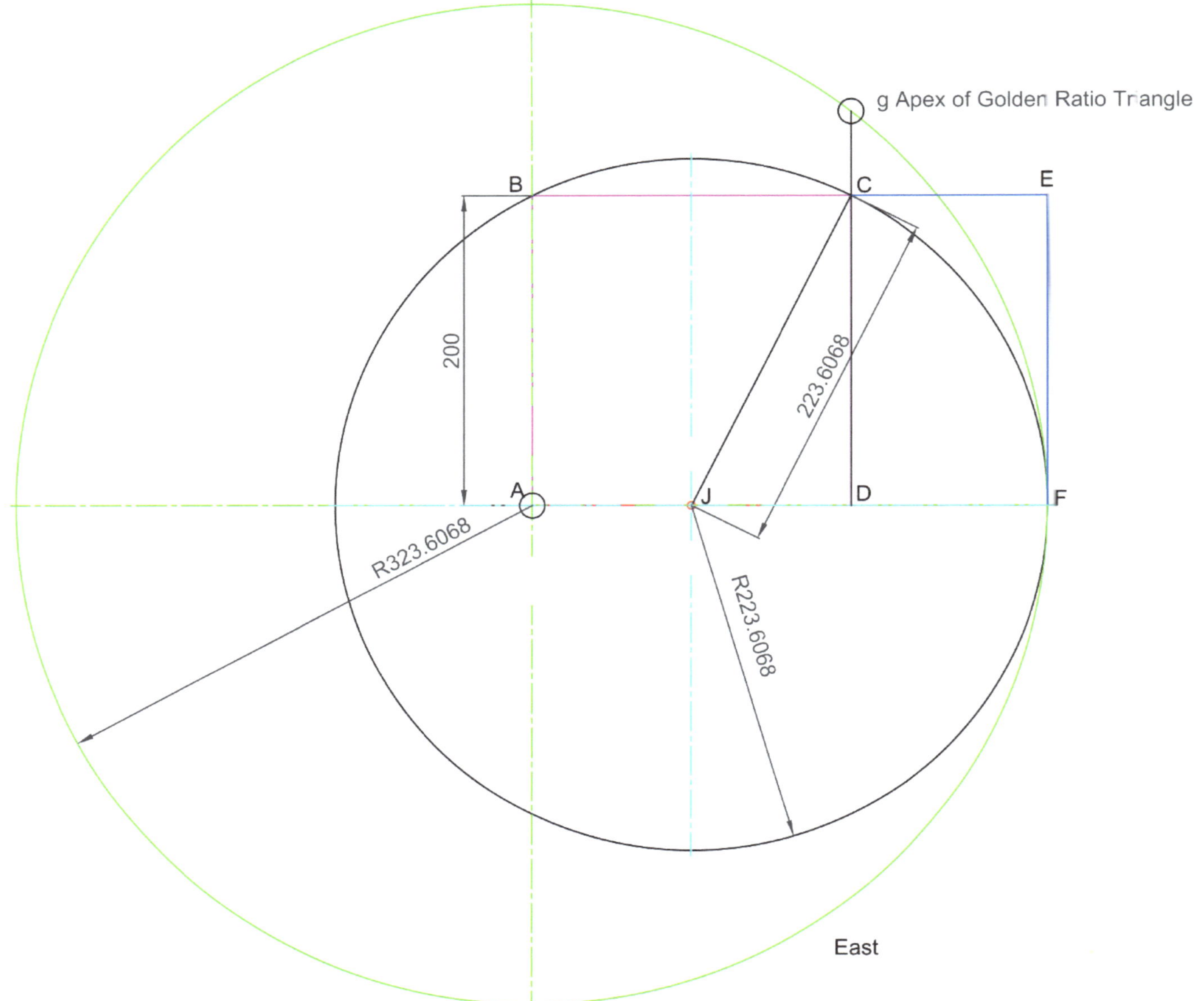

Figure 8 **Finding the Apex for Golden Ratio Triangle**

1. Square = ABCD
2. Right Angle Triangle = JDC
3. Golden Ratio Rectangle = ABEF
4. Black Circle with center J has radius = hypotenuse = JC = 223.6068
5. Green Circle with center A has radius = longer side of Rectangle = AF = 323.6068
6. **Extend** side of Square **DC** to touch Green Circle, this point **g** is **Apex for Golden Ratio Triangle T1u**

7 Draw one Side of the Golden Ratio Triangle

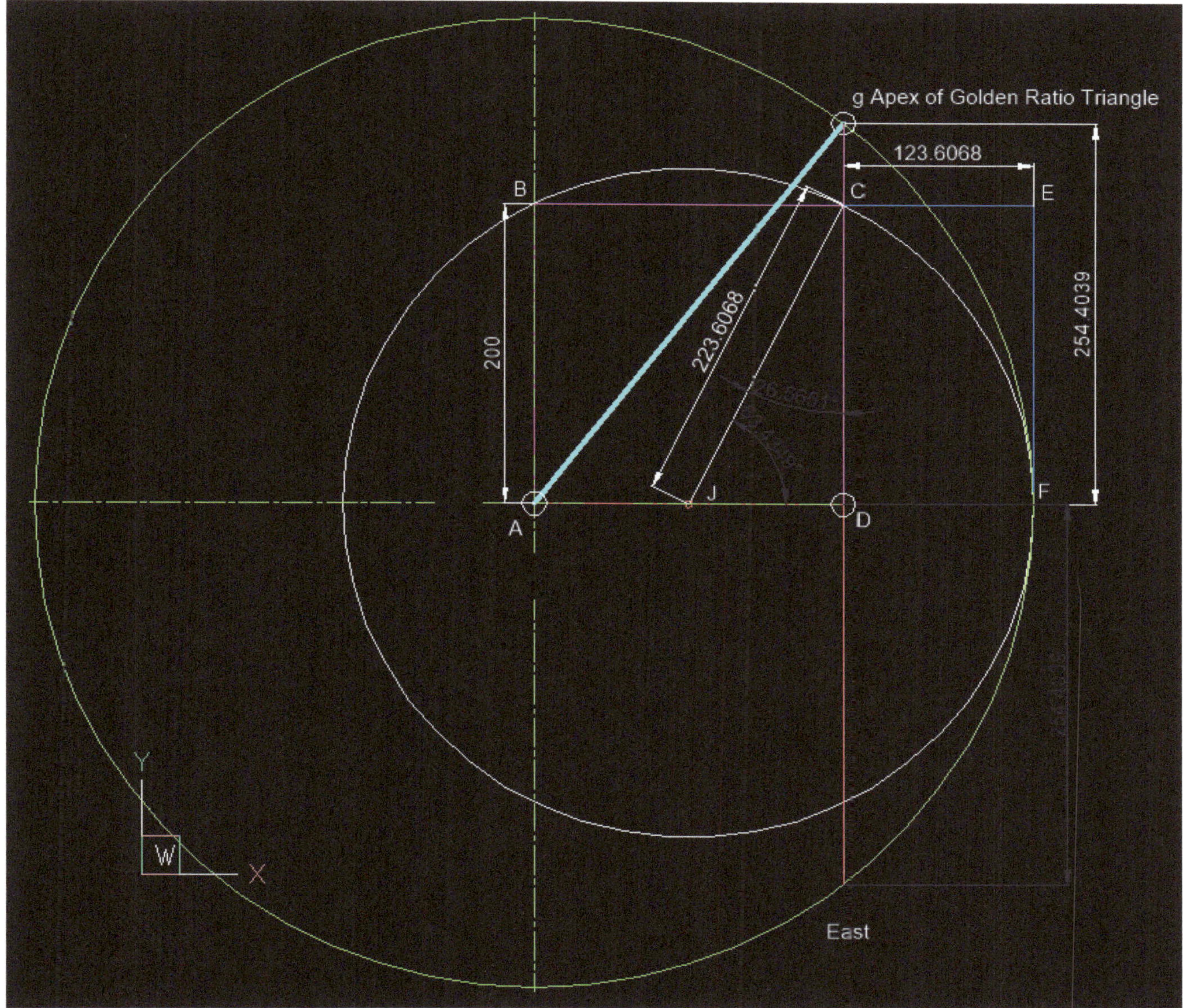

Figure 9 **Draw one Side of the Golden Ratio Triangle**

1. Square = ABCD
2. Right Angle Triangle = JDC
3. Golden Ratio Rectangle = ABEF
4. Black Circle with center J has radius = hypotenuse = JC
5. Green Circle with center A has radius = longer side of Rectangle = AF
6. **Extend** side of Square **DC** to touch Green Circle, this point **g** is Apex for **Golden Ratio Triangle T1u**
7a. Now draw one **Side** of T1u by joining points **Ag**. It gives Angle **gAD** = 51.82729°

23

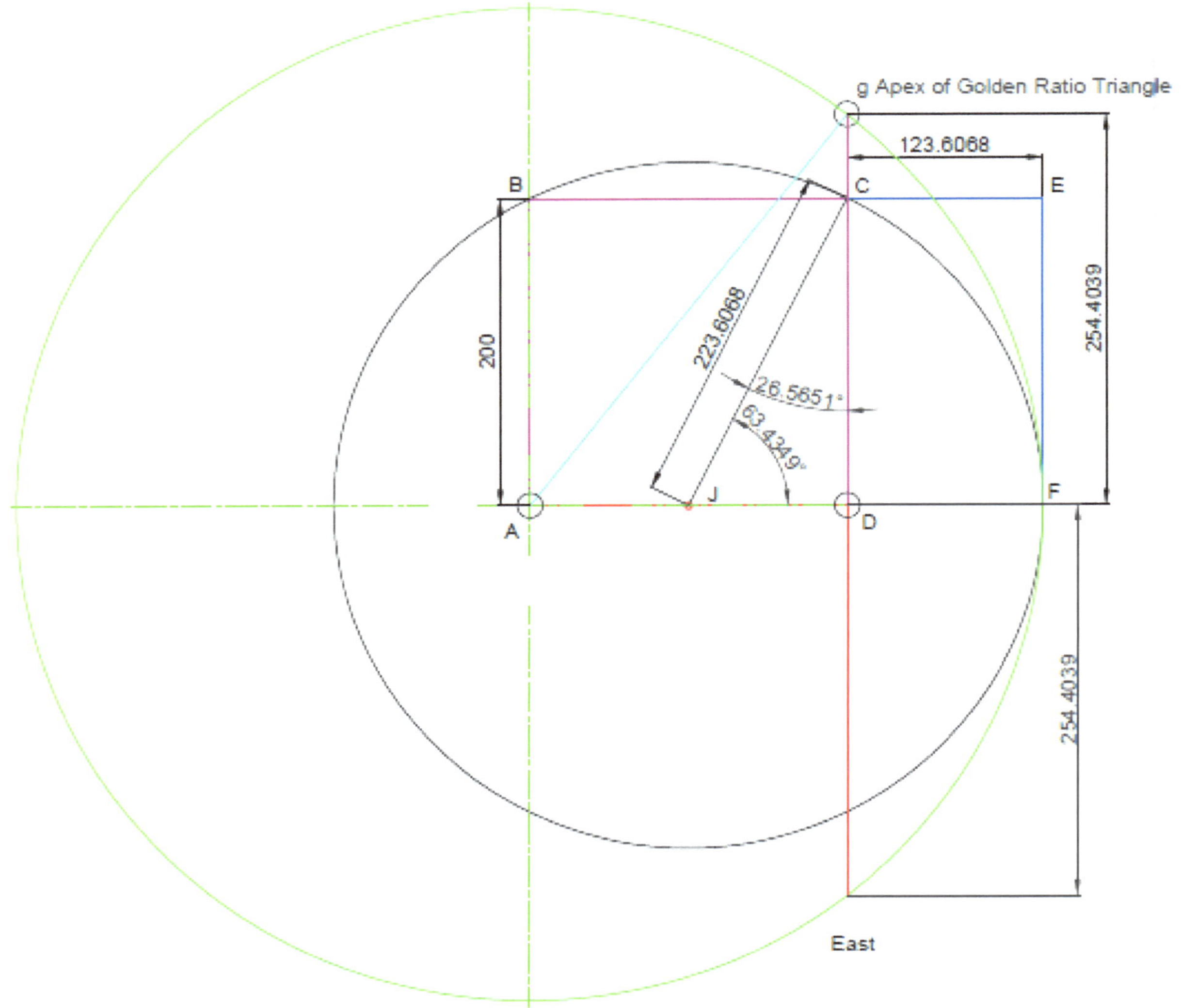

Figure 10 **Draw one Side of the Golden Ratio Triangle (alternate view)**

1. Square = ABCD
2. Right Angle Triangle = JDC
3. Golden Ratio Rectangle = ABEF
4. **Black** Circle with center J has radius = hypotenuse = JC
5. **Green** Circle with center A has radius = longer side of Rectangle = AF
6. **Extend** side of Square **DC** to touch Green Circle, this point **g** is **Apex for Golden Ratio Triangle T1u**
7a. Now draw the **side** of T1u by joining points **Ag**. It gives angle **gAD** = 51.82729°
7b. **Copy and Mirror** the side **Ag** to get the **other** side **gi** for T1u.
7c. Complete the Triangle T1u by joining the points **Ai** to get its **baseline**, thus T1u = **gAi**
7d. T1u dimensions, AF = Ag = gi = 323.6068, Ai = 400. Isosceles Angles = 51.82729° each.

8 Complete the sides of Golden Ratio Triangle

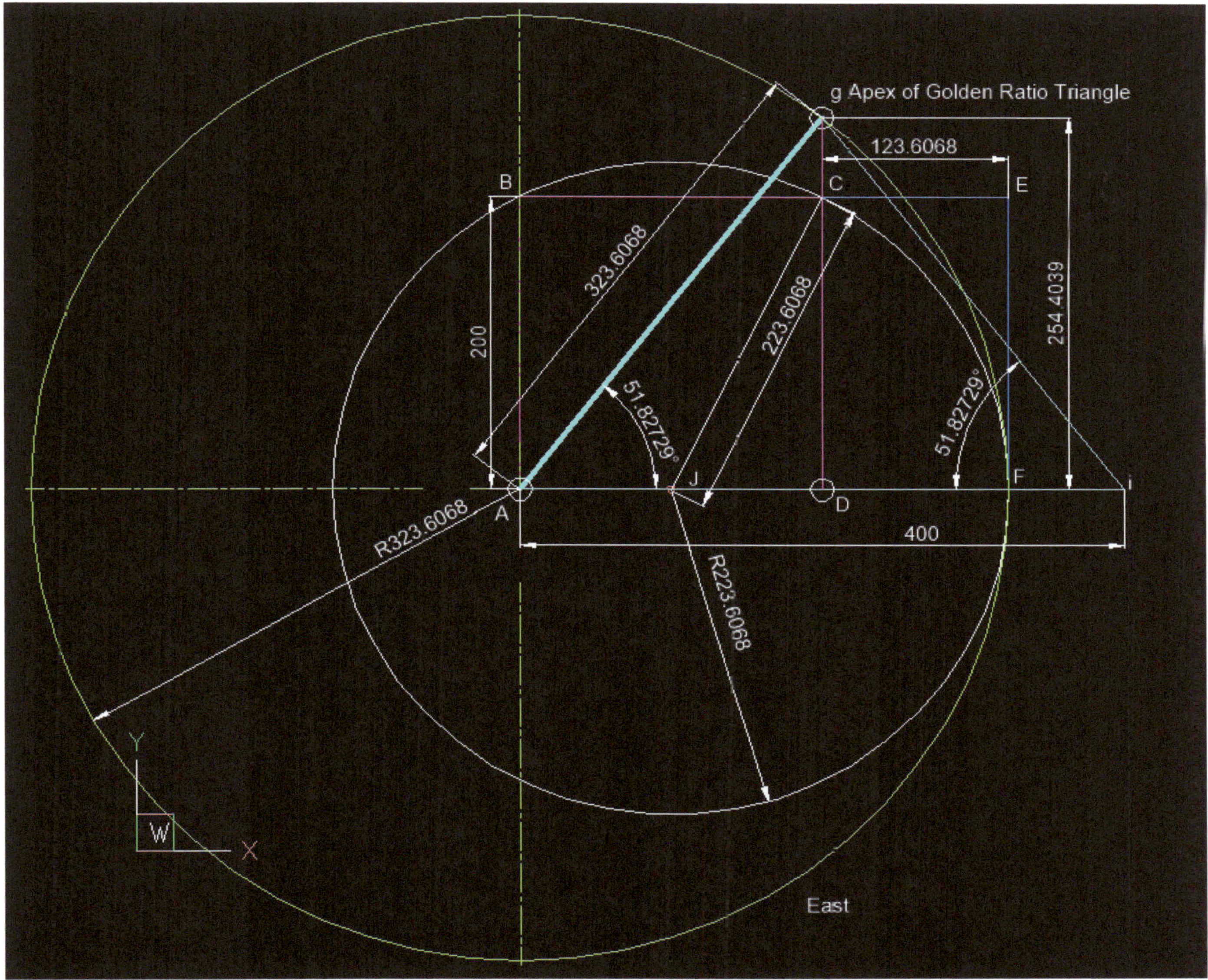

Figure 11 **Complete the Golden Ratio Triangle**

1. Square = ABCD
2. Right Angle Triangle = JDC
3. Golden Ratio Rectangle = ABEF
4. Black Circle with center J has radius = hypotenuse = JC
5. Green Circle with center A has radius = longer side of Rectangle = AF
6. **Extend** side of Square **DC** to touch Green Circle, this point **g** is Apex for Golden Ratio Triangle T1u
7a. Now draw the **side** of T1u by joining points **Ag**. It gives angle **gAD** = 51.82729°
7b. **Copy and Mirror** the side **Ag** to get the **other** side **gi** for T1u.
7c. Complete the Triangle T1u by joining the points **Ai** to get its **baseline**, thus T1u = **gAi**
7d. T1u dimensions, AF = Ag = gi = 323.6068, Ai = 400. Isosceles Angles = 51.82729° each.

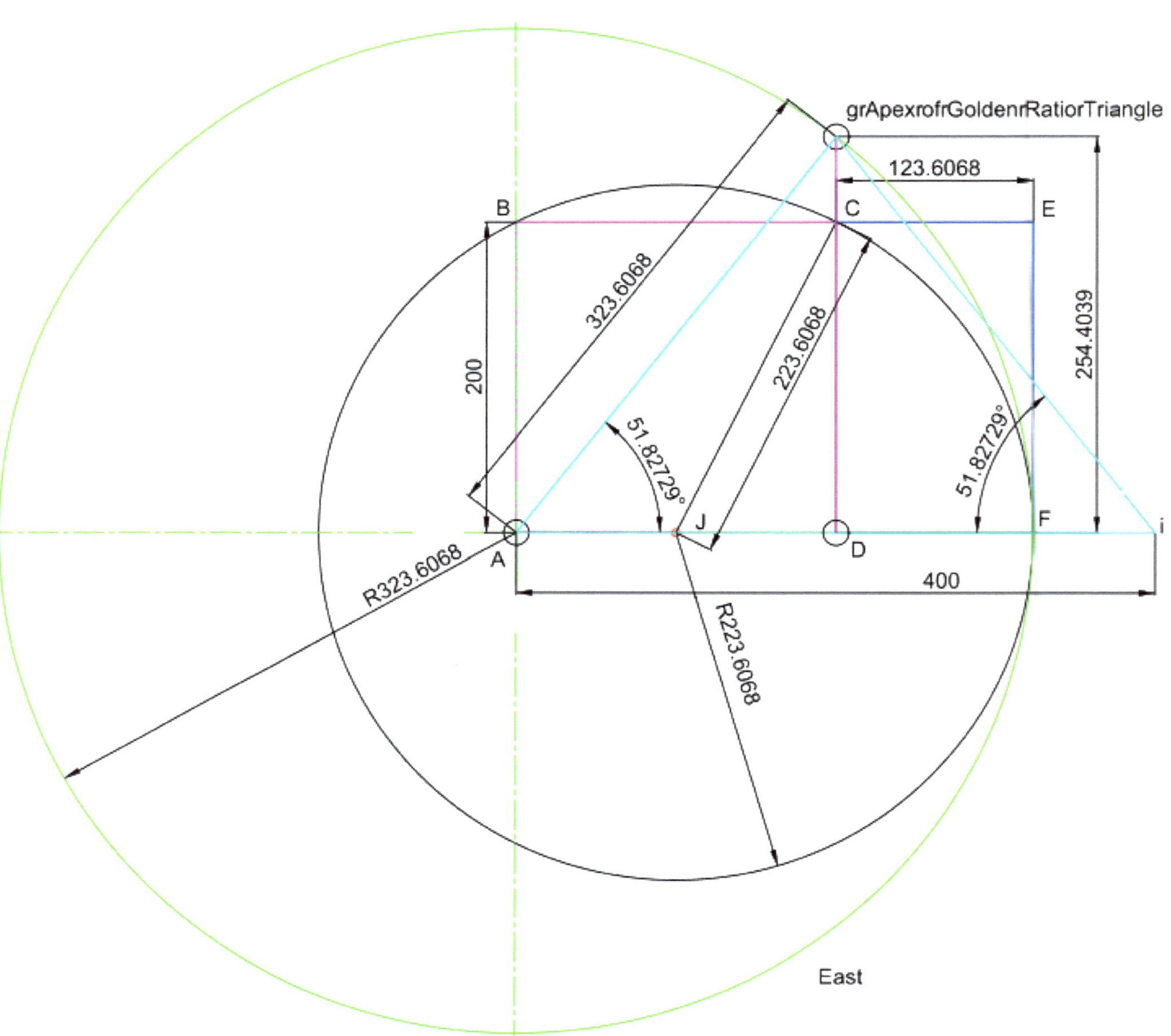

grApexrofrGoldenrRatiorTriangle
123.6068
B
C
E
323.6068
200
223.6068
254.4039
51.82729°
51.82729°
J
F
i
R323.6068
A
D
R223.6068
400
East

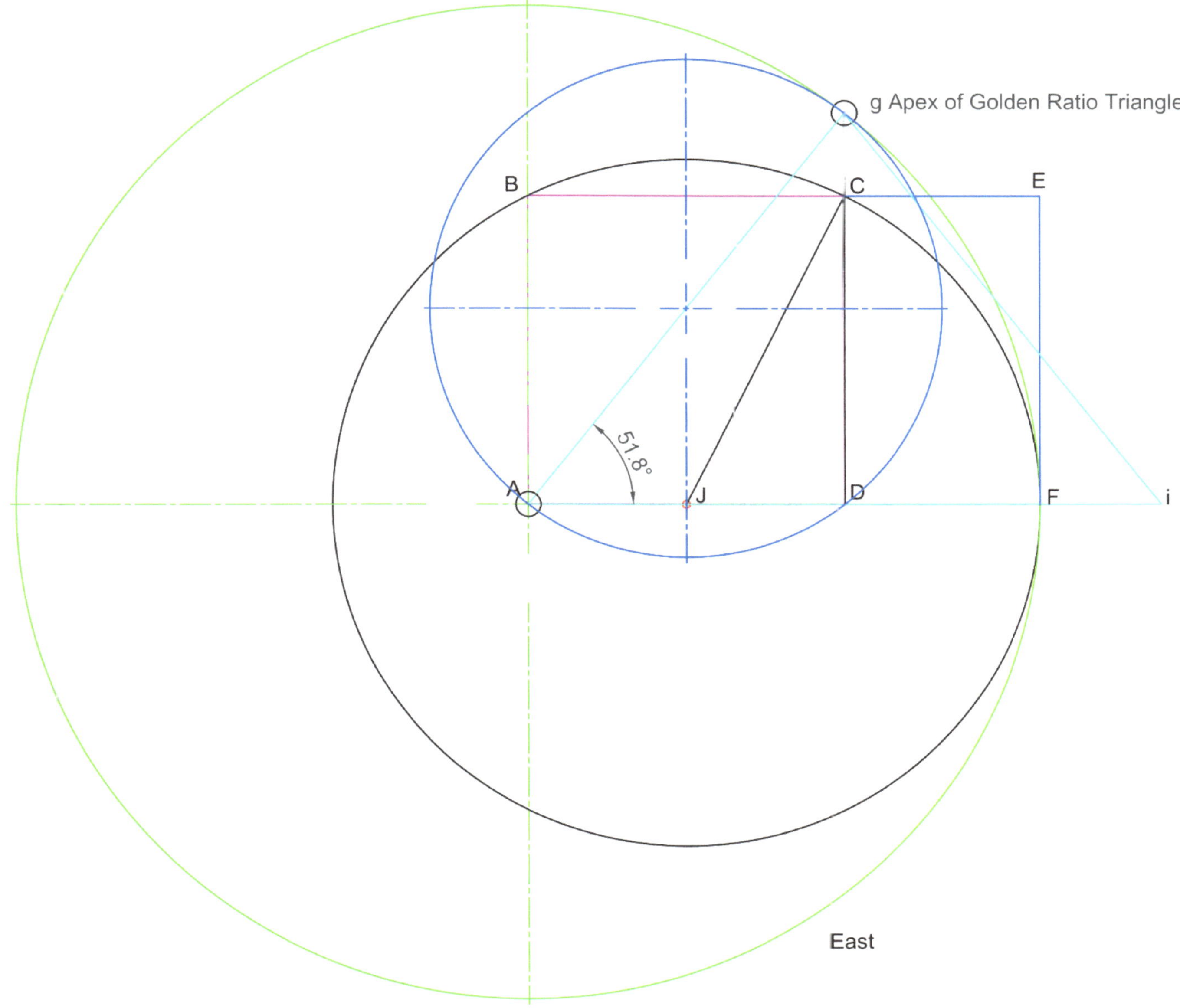

Figure 12 **Finding the Side Center to Bisect Side of Golden Ratio Triangle**

1. Square = ABCD
2. Right Angle Triangle = JDC
3. Golden Ratio Rectangle = ABEF
4. Black Circle with center J has radius = hypotenuse = JC
5. Green Circle with center A has radius = longer side of Rectangle = AF
6. **Extend** side of Square **DC** to touch Green Circle, this point **g** is **Apex** for **Golden Ratio Triangle T1u**
7. Complete drawing all sides of Triangle T1u
8. Blue Circle with **diameter** Ag to find **centerpoint** of side Ag of Triangle T1u.

10 Using Side centerpoint Bisect the T1u Side

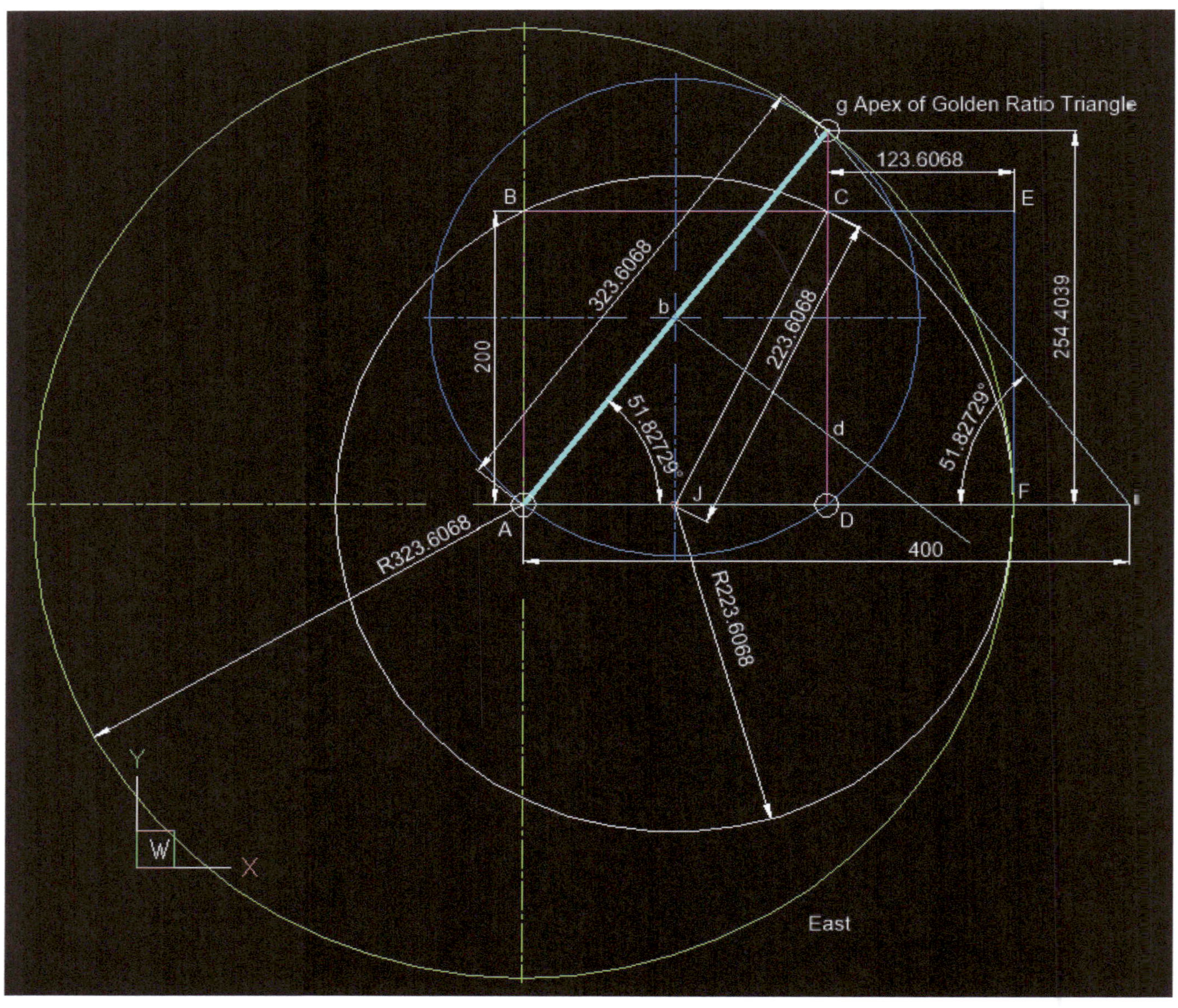

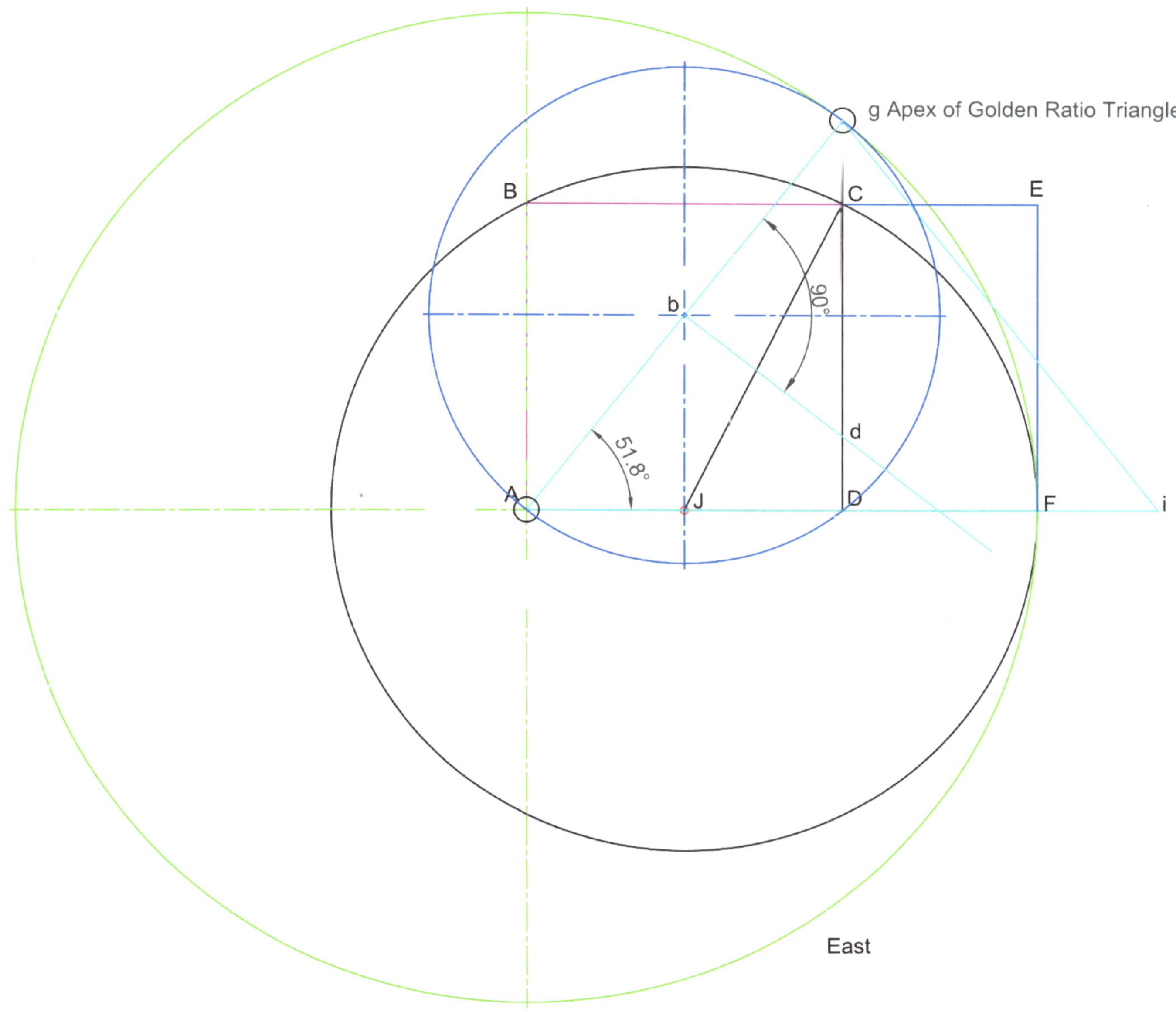

Figure 13 **Bisect Sides of Triangle T1u to get Center Point of Circumscribe Circle**

1. Square = ABCD
2. Right Angle Triangle = JDC
3. Golden Ratio Rectangle = ABEF
4. Black Circle with center J has radius = hypotenuse = JC
5. Green Circle with center A has radius = longer side of Rectangle = AF
6. **Extend** side of Square **DC** to touch Green Circle, this point **g** is Apex **for Golden Ratio Triangle T1u**
7. Complete drawing all sides of Triangle T1u
8a. Blue Circle with **diameter** Ag to find **centerpoint** of side Ag of Triangle T1u.
8b. Vertical CD of right angle triangle JCD already **bisects the base** of Triangle T1u.
9. **Bisect** the T1u side Ag to obtain Point **d** on **Vertical** of Right Angle Triangle JDC.

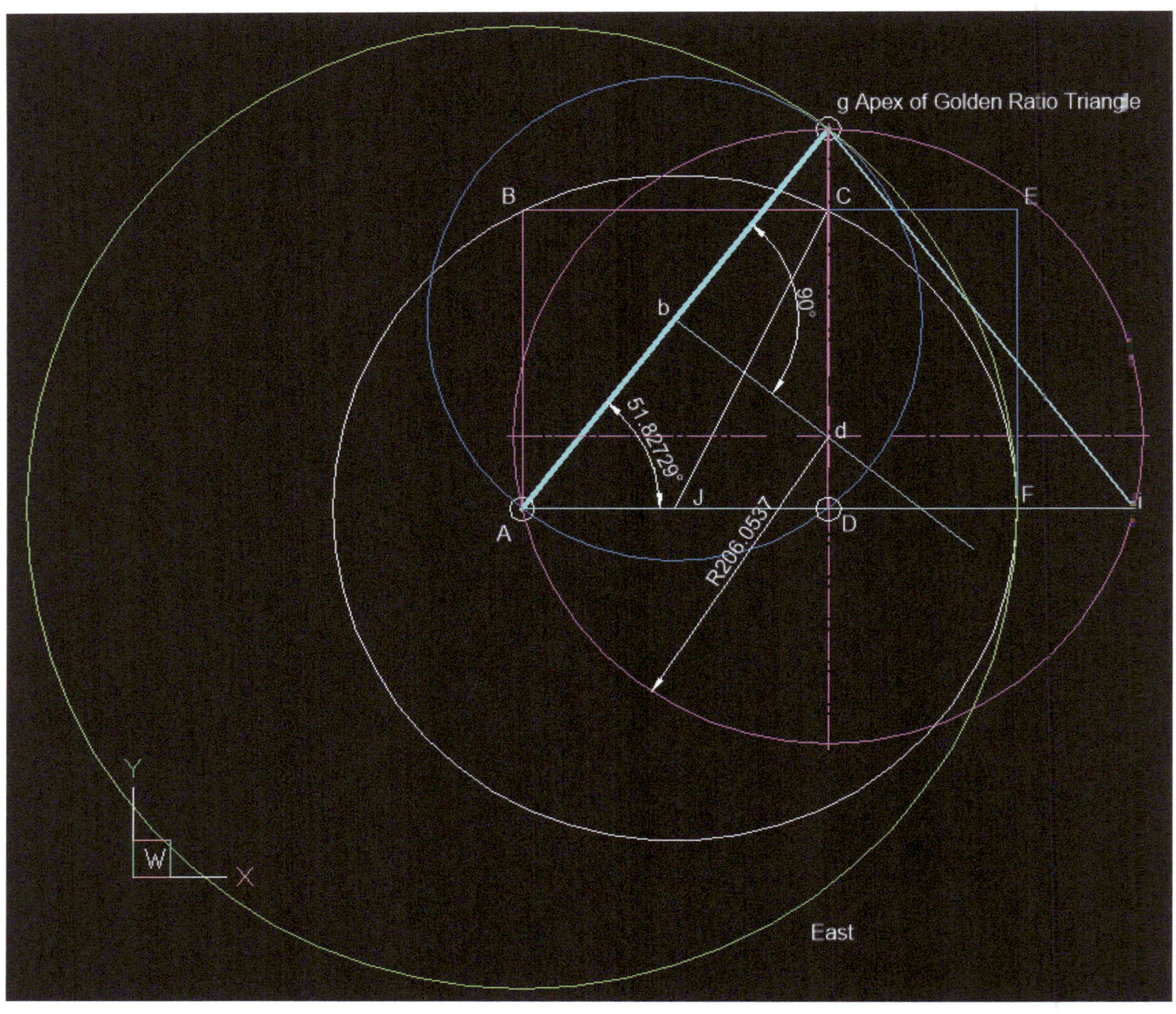

g Apex of Golden Ratio Triangle
B
C
E
b
90°
d
51.82729°
J
R206.0537
A
D
F
i
Y
W
X
East

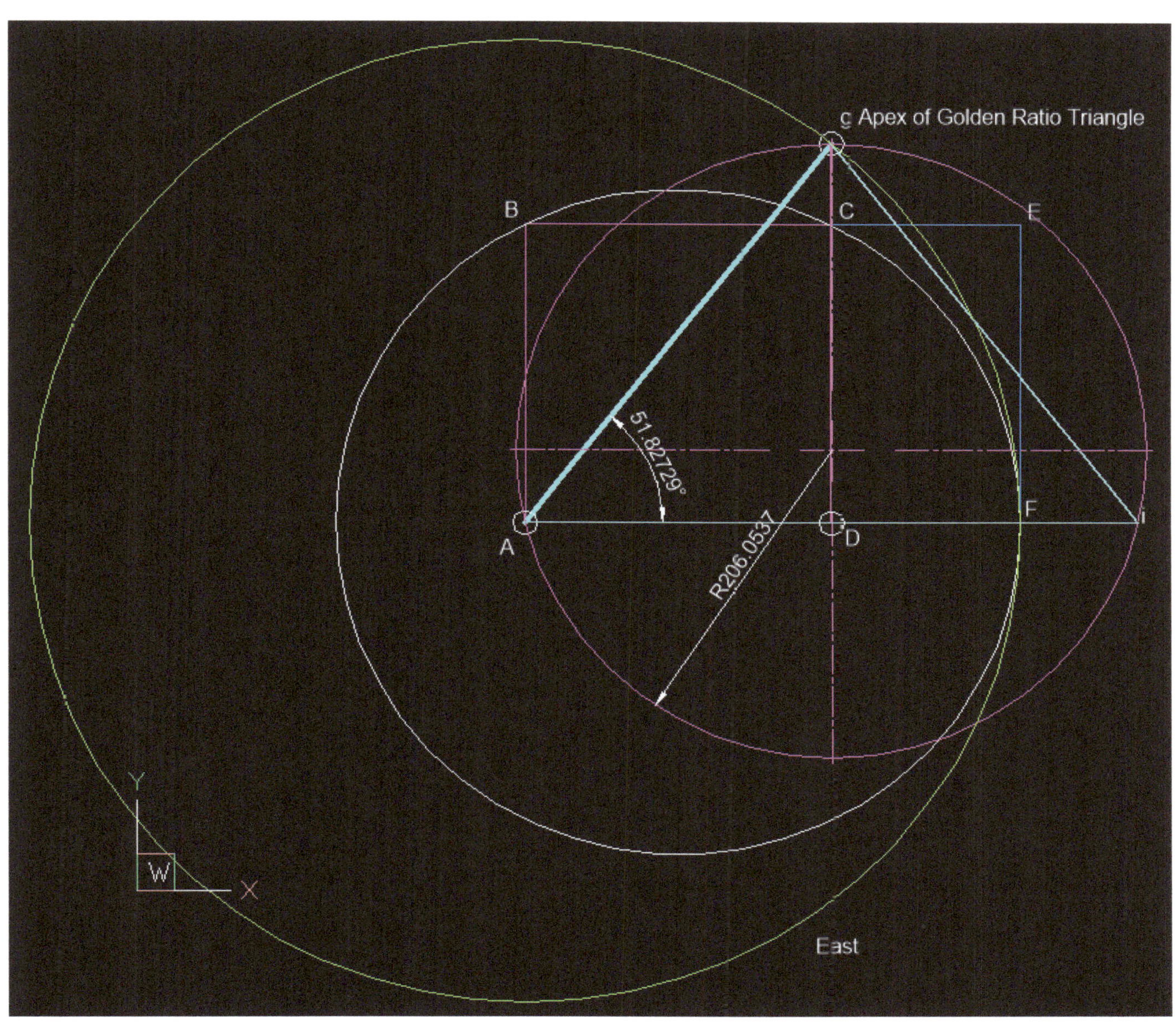

Apex of Golden Ratio Triangle
B
C
E
51.82729°
R206.0537
A
D
F
Y
W
X
East

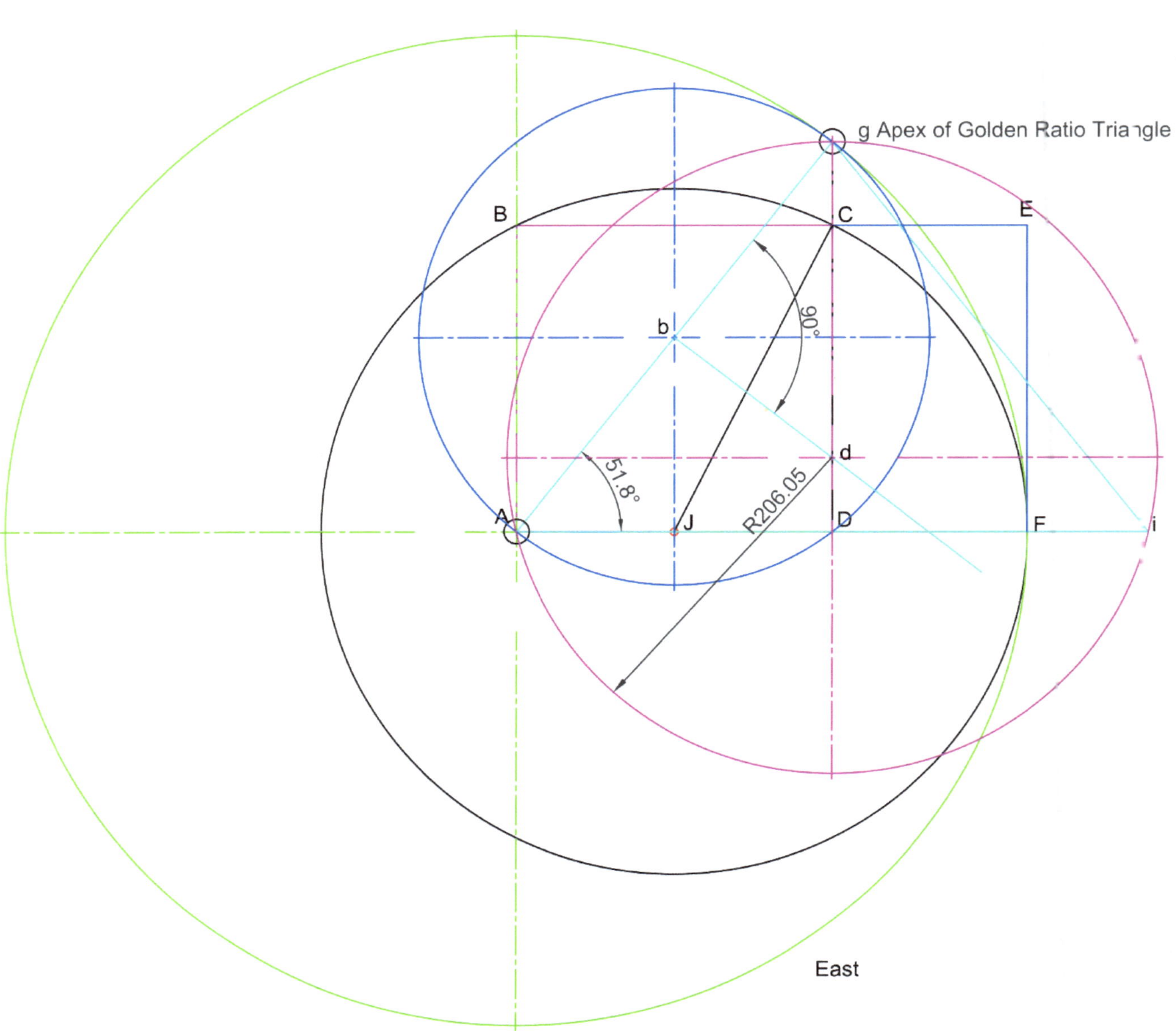

g Apex of Golden Ratio Triangle
B
C
E
b
90°
51.8°
A
J
d
R206.05
D
F
i
East

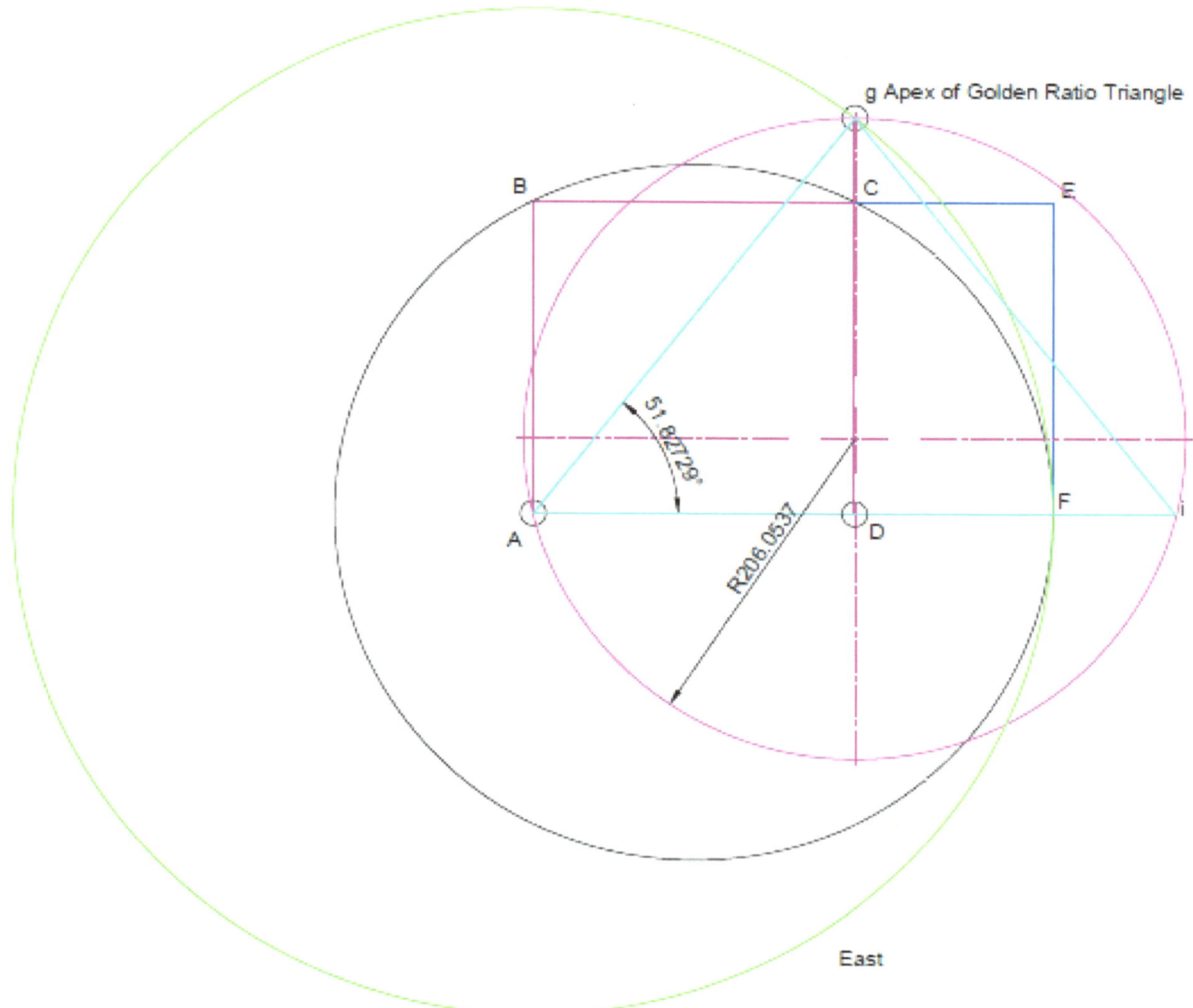

Figure 14 **Draw a Circle to Circumscribe the Golden Ratio Triangle**

1. Square = ABCD
2. Right Angle Triangle = JDC
3. Golden Ratio Rectangle = ABEF
4. **Black** Circle with center J has radius = hypotenuse = JC
5. **Green** Circle with center A has radius = longer side of Rectangle = AF
6. **Extend** side of Square **DC** to touch Green Circle, this point **g** is **Apex** for **Golden Ratio Triangle T1u**
7. Complete drawing all sides of Triangle T1u
8. **Blue** Circle with **diameter** Ag to find **centerpoint** of side Ag of Triangle T1u.
9. **Bisect** the T1u side Ag to **obtain** Point **d** on **Vertical** of Right Angle Triangle JDC.
10. Using Point d as **center**, draw **Magenta** 4th Circle to **circumscribe** Triangle T1u. **Radius 206.0537**

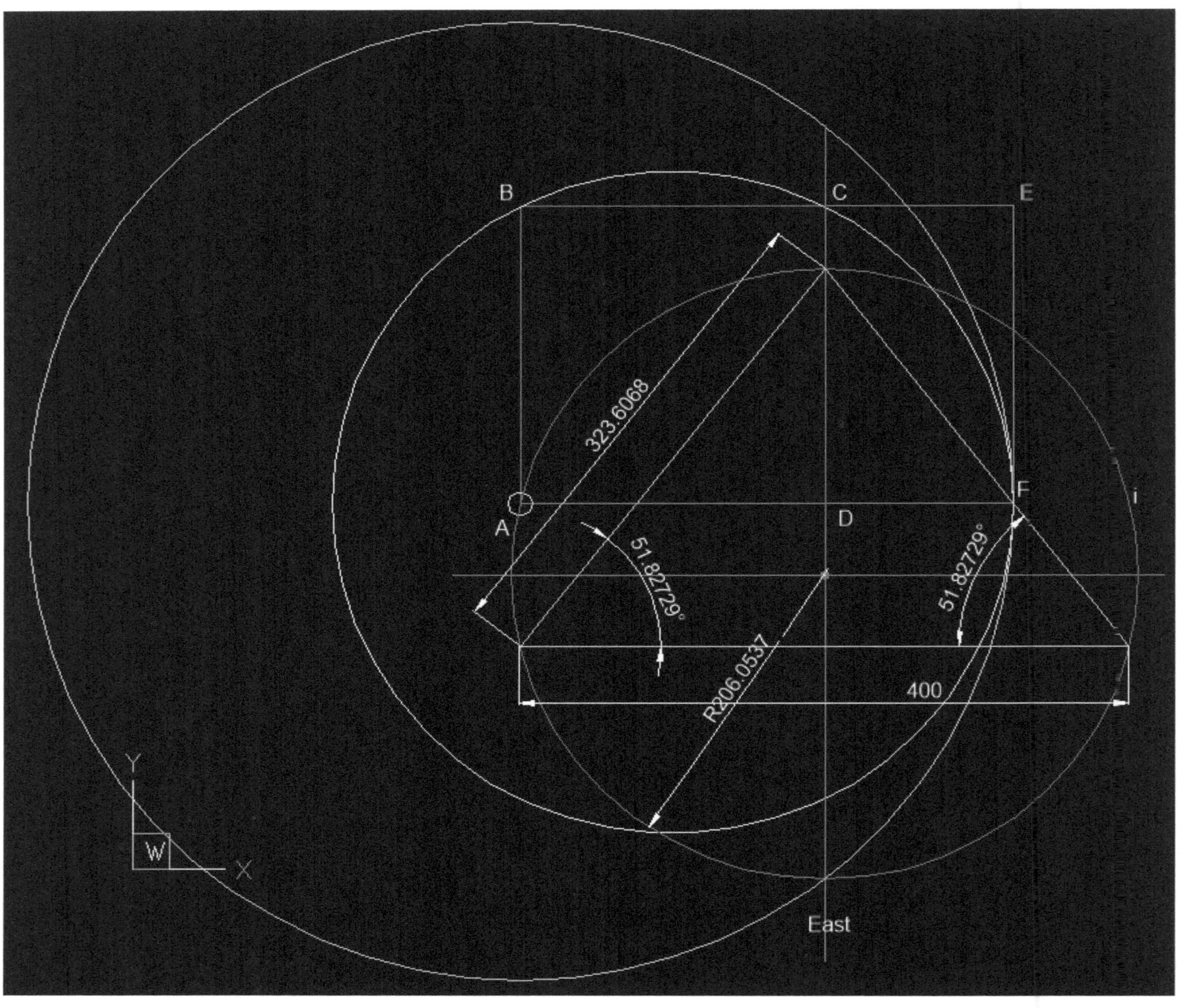

B
C
E
323.6068
51.82729°
51.82729°
R206.0537
400
A
D
F
i
Y
W
X
East

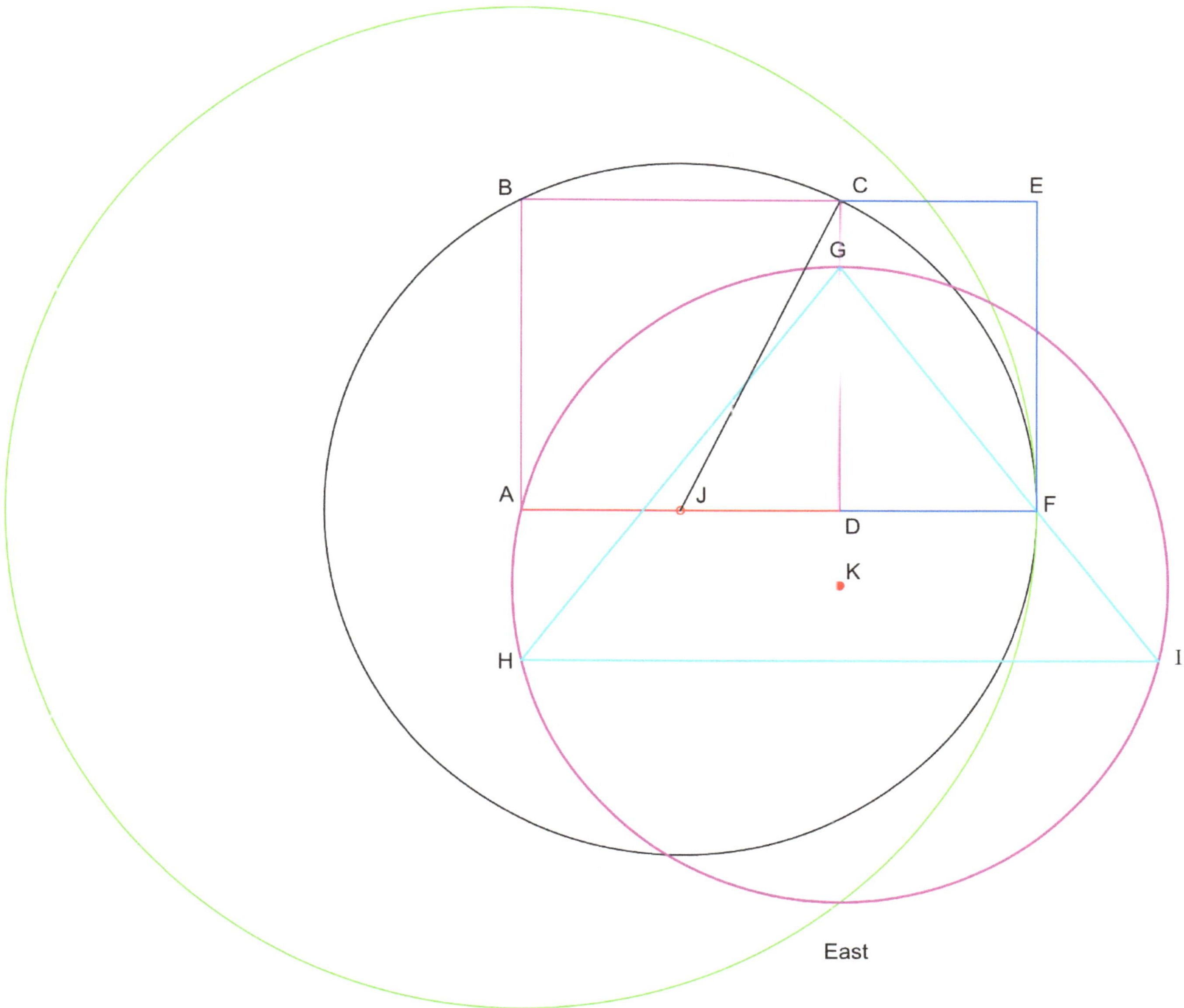

Figure 15 **Square Rectangle Circles Golden Ratio Triangle**

1. Square = ABCD
2. Right Angle Triangle = JDC
3. Golden Ratio Rectangle = ABEF
4. Circle with center J has radius = hypotenuse = JC
5. Circle with center A has radius = longer side of Rectangle = AF
6. Golden Ratio Triangle T1u = GHI
7. **Circle** circumscribing Golden Ratio Triangle with center K has radius = **Radius 206.05**
8. We **shift** Triangle T1u and its circumscribing Circle **arbitrarily**, simply for **aesthetics**.
(Note: This step can be skipped. We can continue drawing the Sri Yantra without shifting.)

13 Draw Cross Hair Center mark

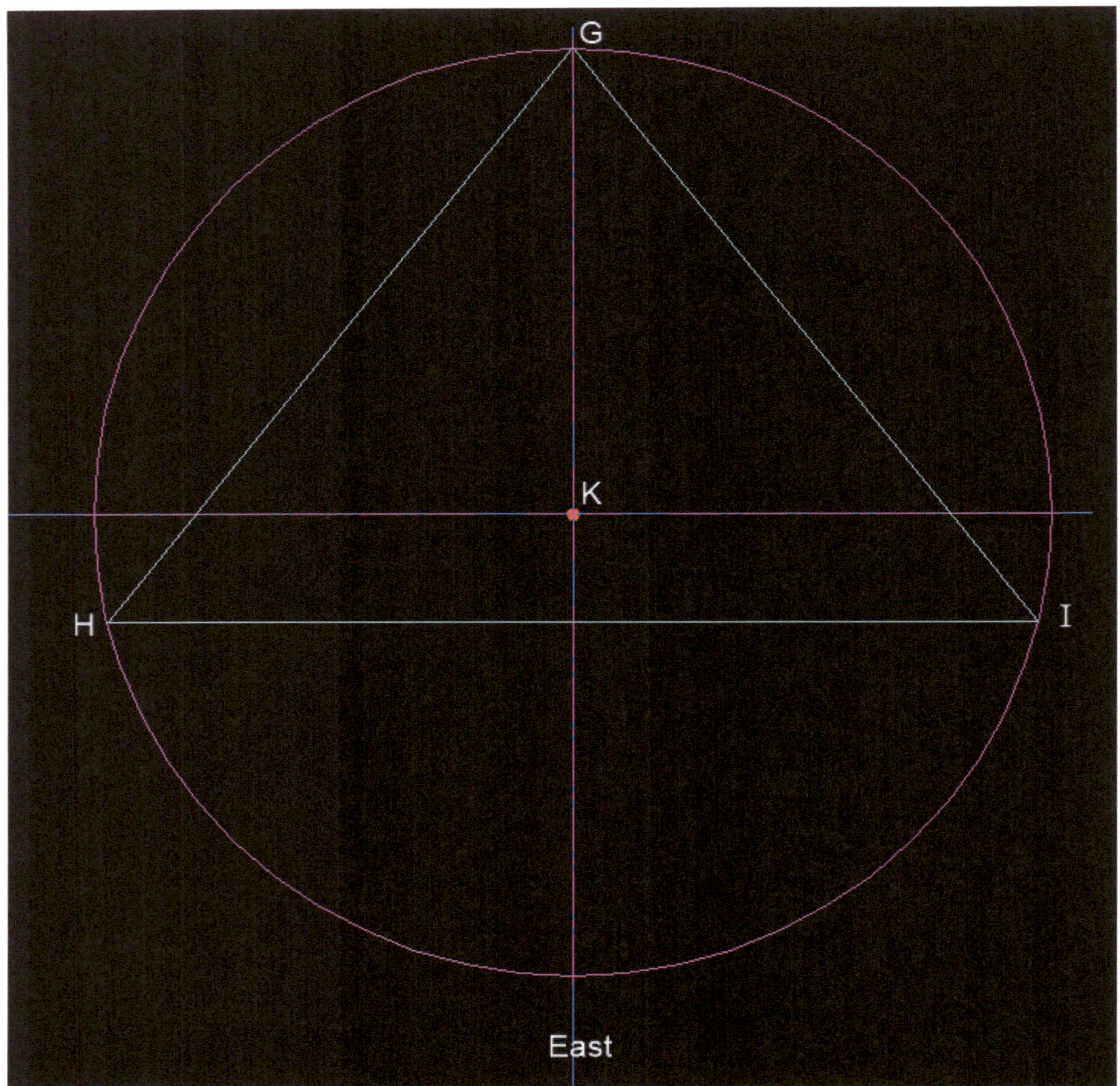

Figure 16 **Cross Hair**

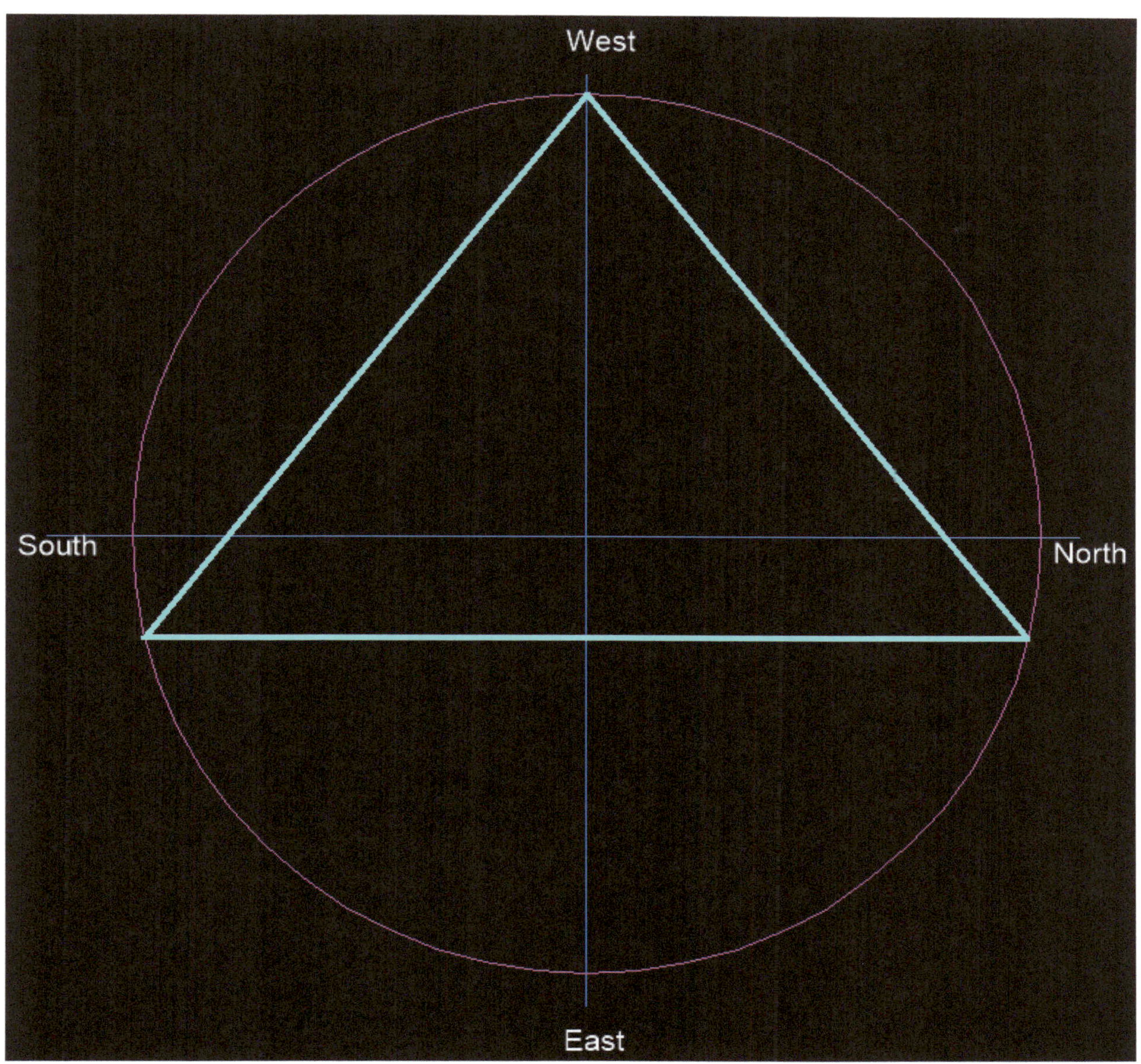

West
37
South
North
East

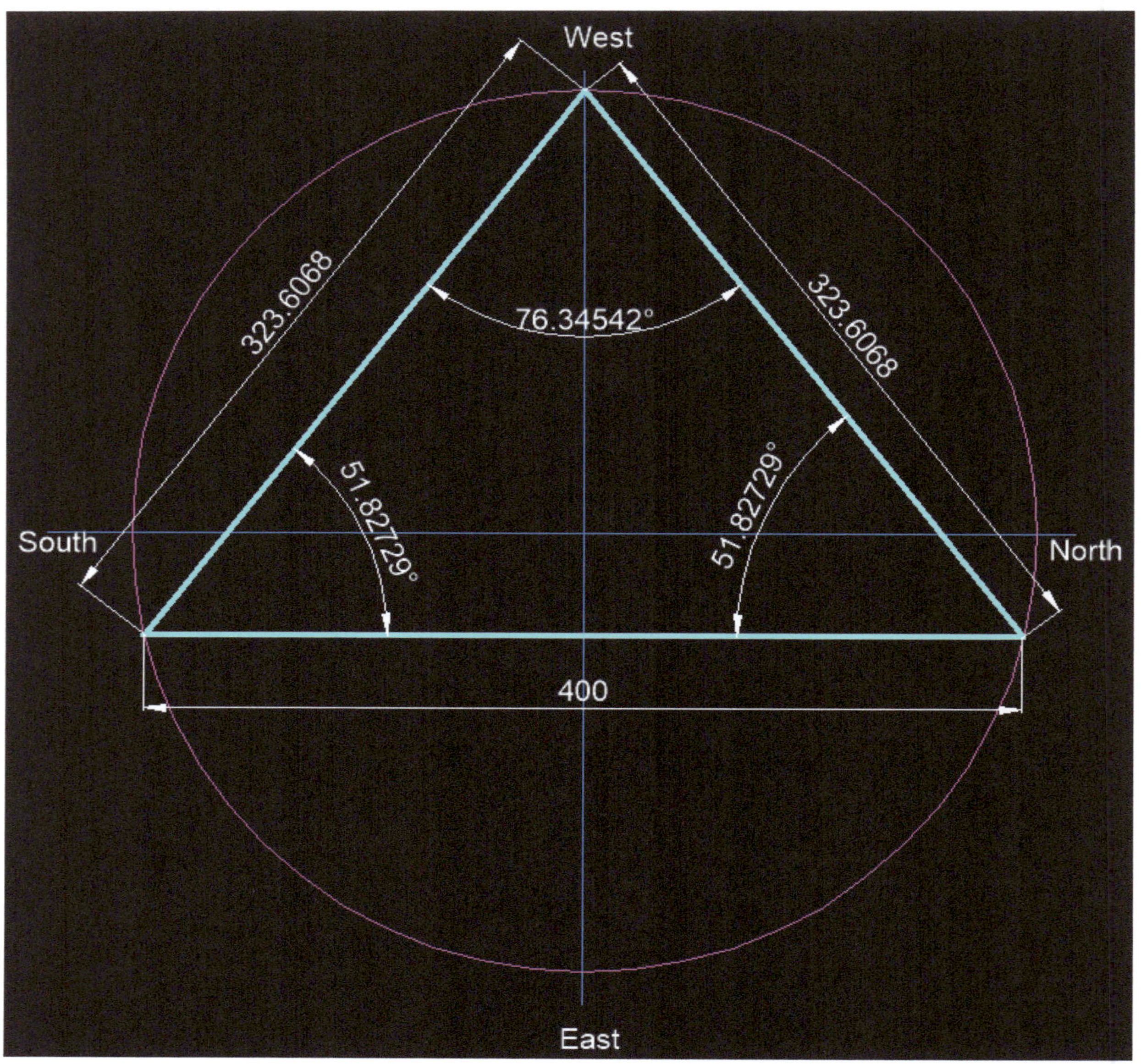

West
South
North
East
323.6068
323.6068
76.34542°
51.82729°
51.82729°
400

14 Draw a 3 4 5 Right Angle Triangle Downward Apex

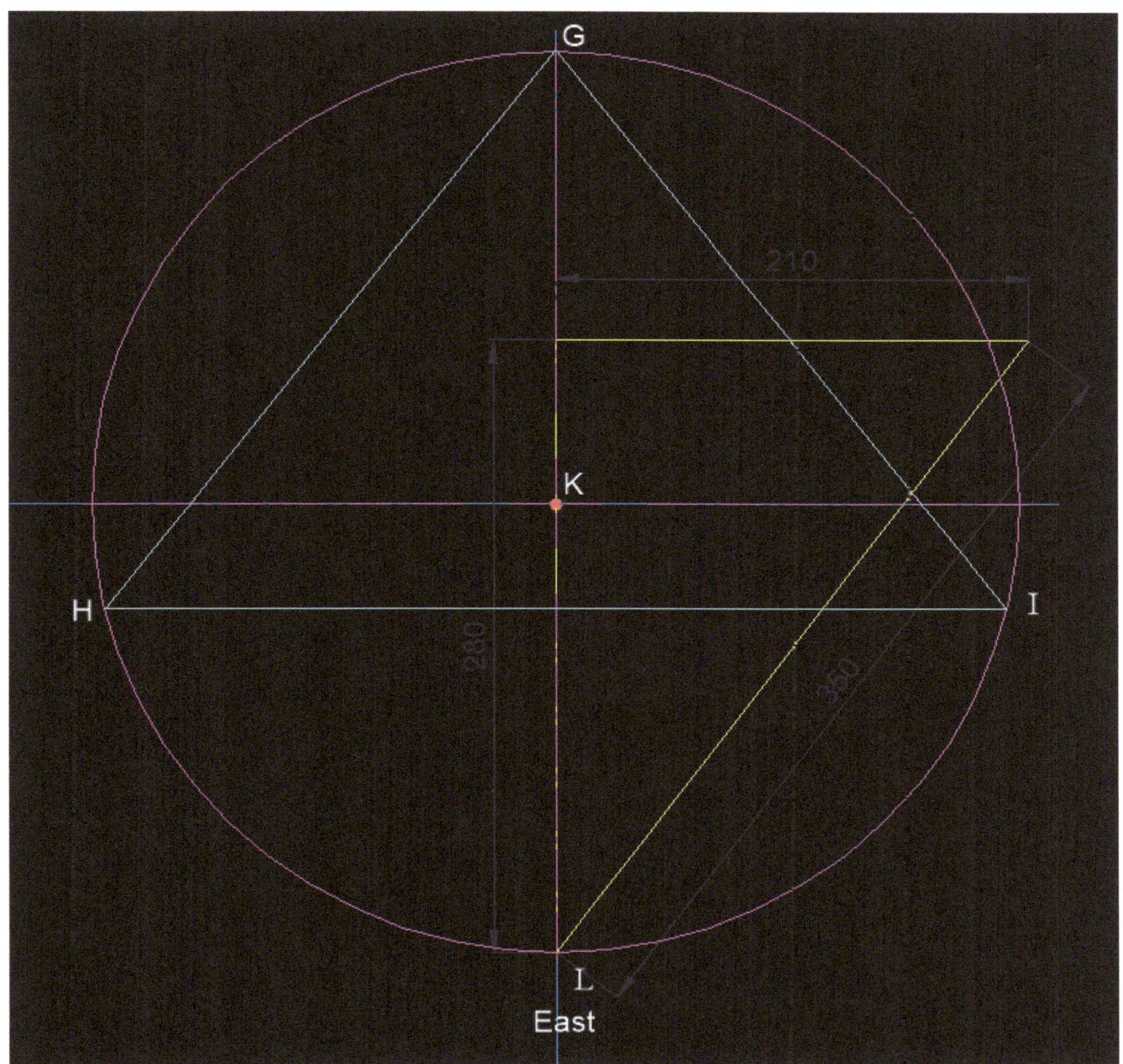

Figure 17 **3 4 5 Right Angle Triangle with Downward Apex**

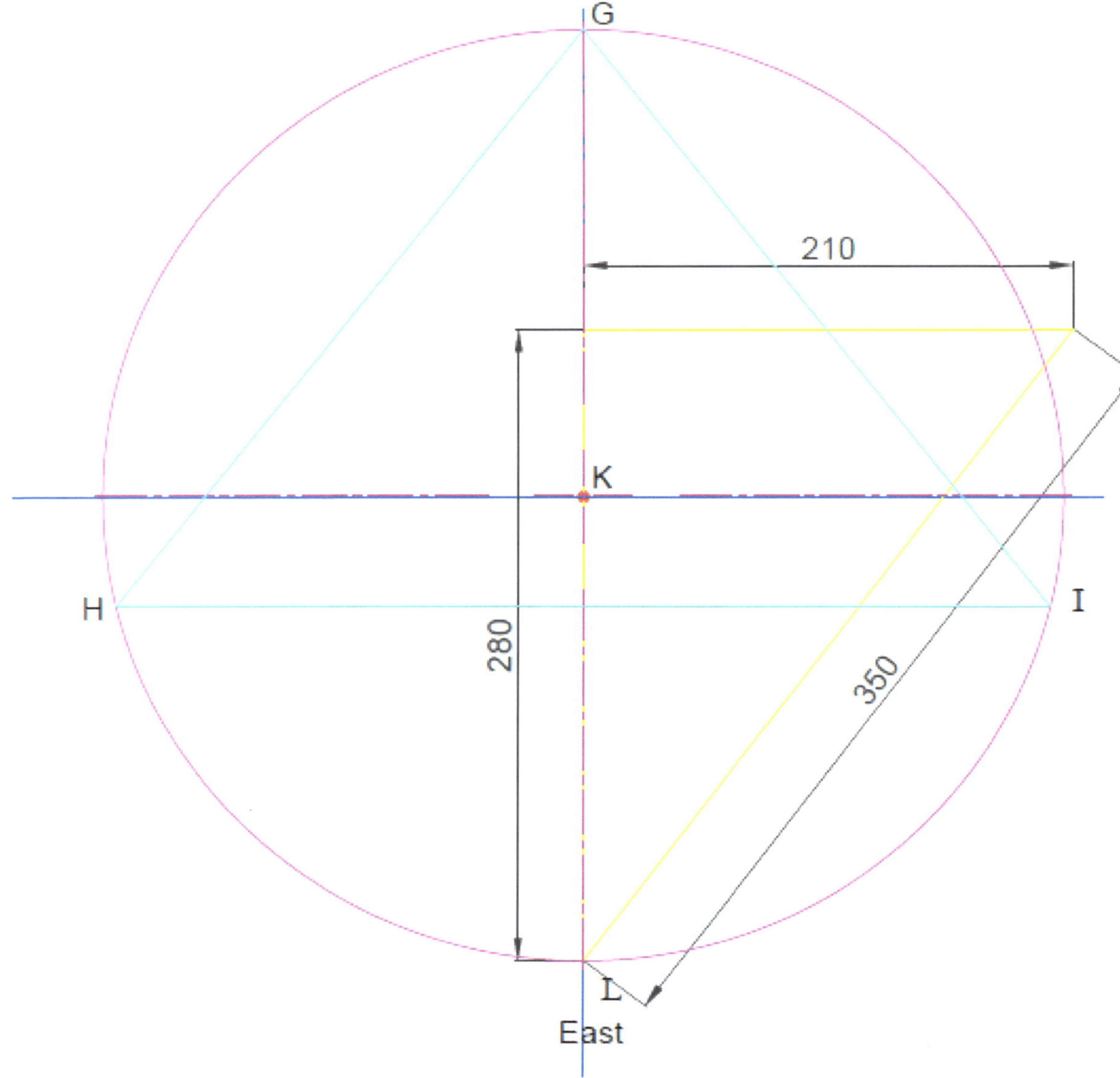

Figure 18 **3 4 5 Right Angle Triangle Multiplier is 7**

1. Draw a 3 4 5 Triangle with Apex at Point L, where crosshair touches the circle
2. The sides of this triangle are in the ratio 3:4:5 = baseline:vertical:hypotenuse
3. We take the base = 3x7 = 210, vertical = 4x7 = 280, hypotenuse = 5x7 = 350.
4. Note regarding multiplier:
    a) We need to have a **whole** number multiplier.
    b) The Vertical so made must be **above** the Center K of Circle.
    c) The Base line must extend **beyond** the Circle.
    The **smallest** whole number that satisfies a) b) c) simultaneously is **7**.

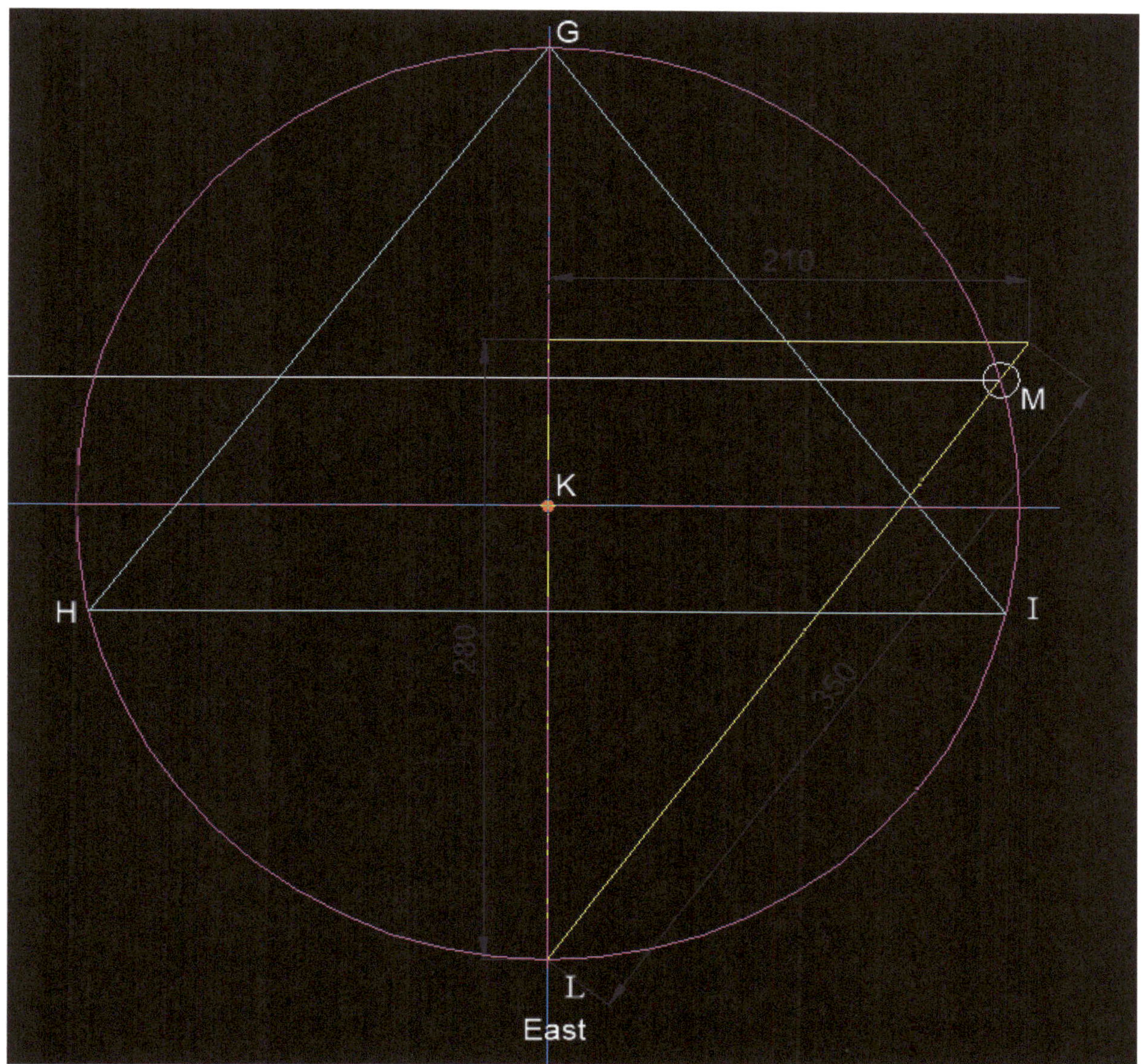

Figure 19 **Side Marker for T1d**

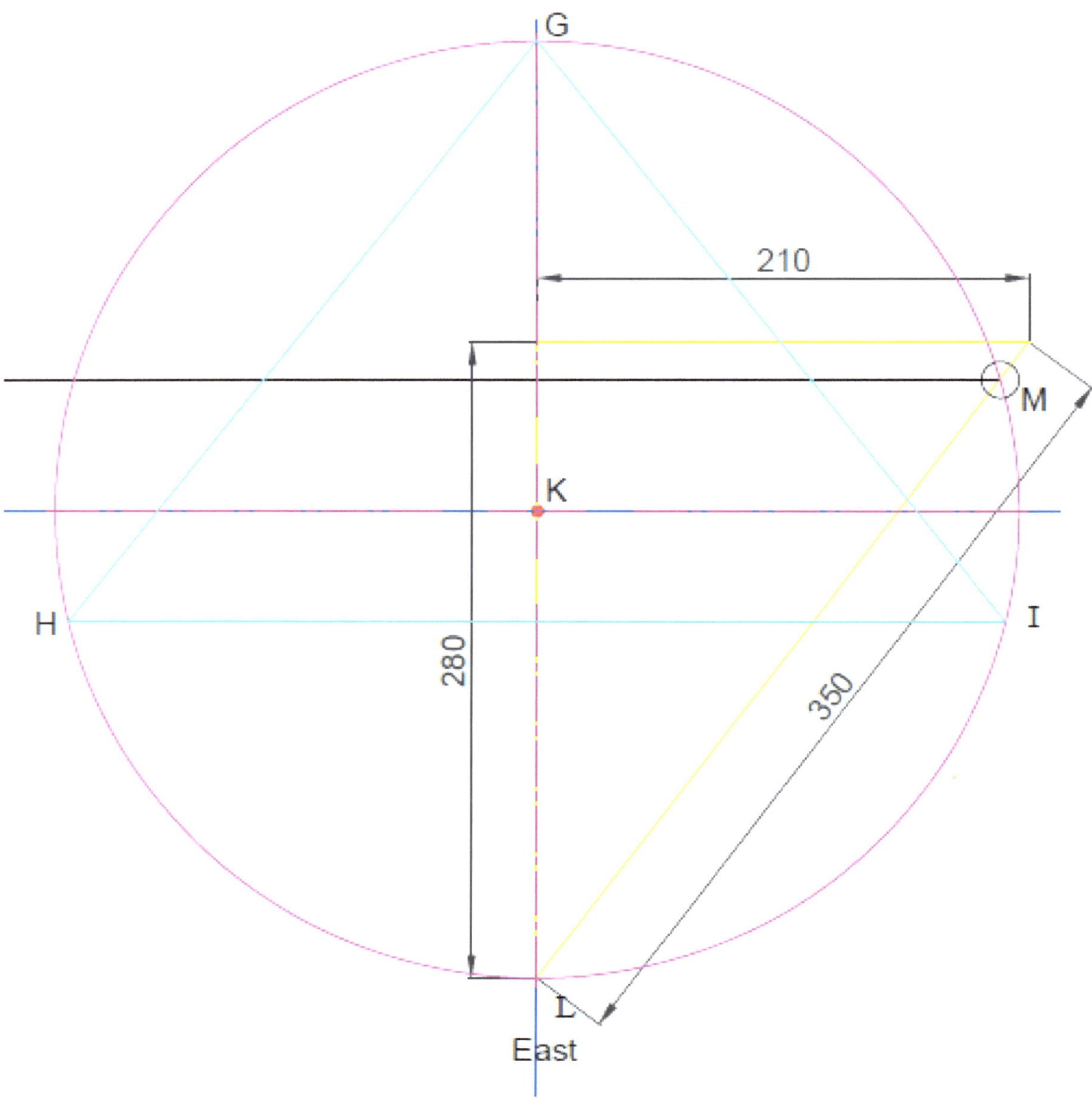

1. Draw a 3 4 5 Triangle with Apex at Point L, where crosshair touches the circle
2. Since our multiplier is 7, we get 210 280 350 Right Angle Triangle.
3. Note the point M where this triangle touches the circle.
4. From point M draw a horizontal to touch other side of circle. This gives us the base for T1d.
5. Using the Base points and the Apex at L, draw the sides of T1d.
6. Thus we obtain T1d, the largest downward apex triangle.

16 Complete Largest Downward Apex Triangle T1d

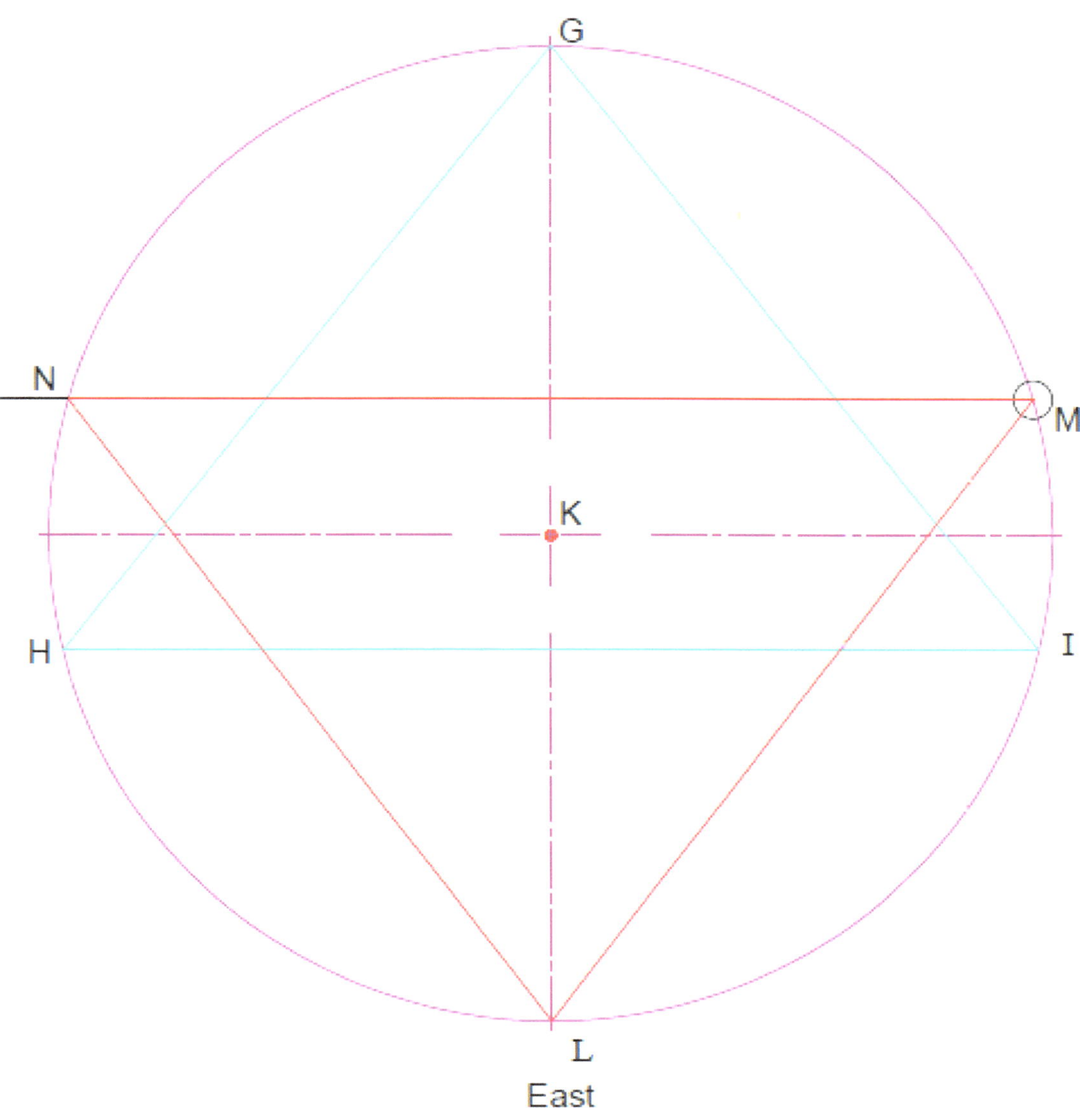

1. Draw a 3 4 5 Triangle with Apex at Point L, where crosshair touches the circle
2. Since our multiplier is 7, we get 210 280 350 Right Angle Triangle.
3. Note the point M where this triangle touches the circle.
4. From point M draw a horizontal to touch other side of circle at N. This gives us the base for T1d.
5. Using the Base points and the Apex at L, draw the sides of T1d, NL and ML.
6. Thus we obtain T1d, the largest downward apex triangle.

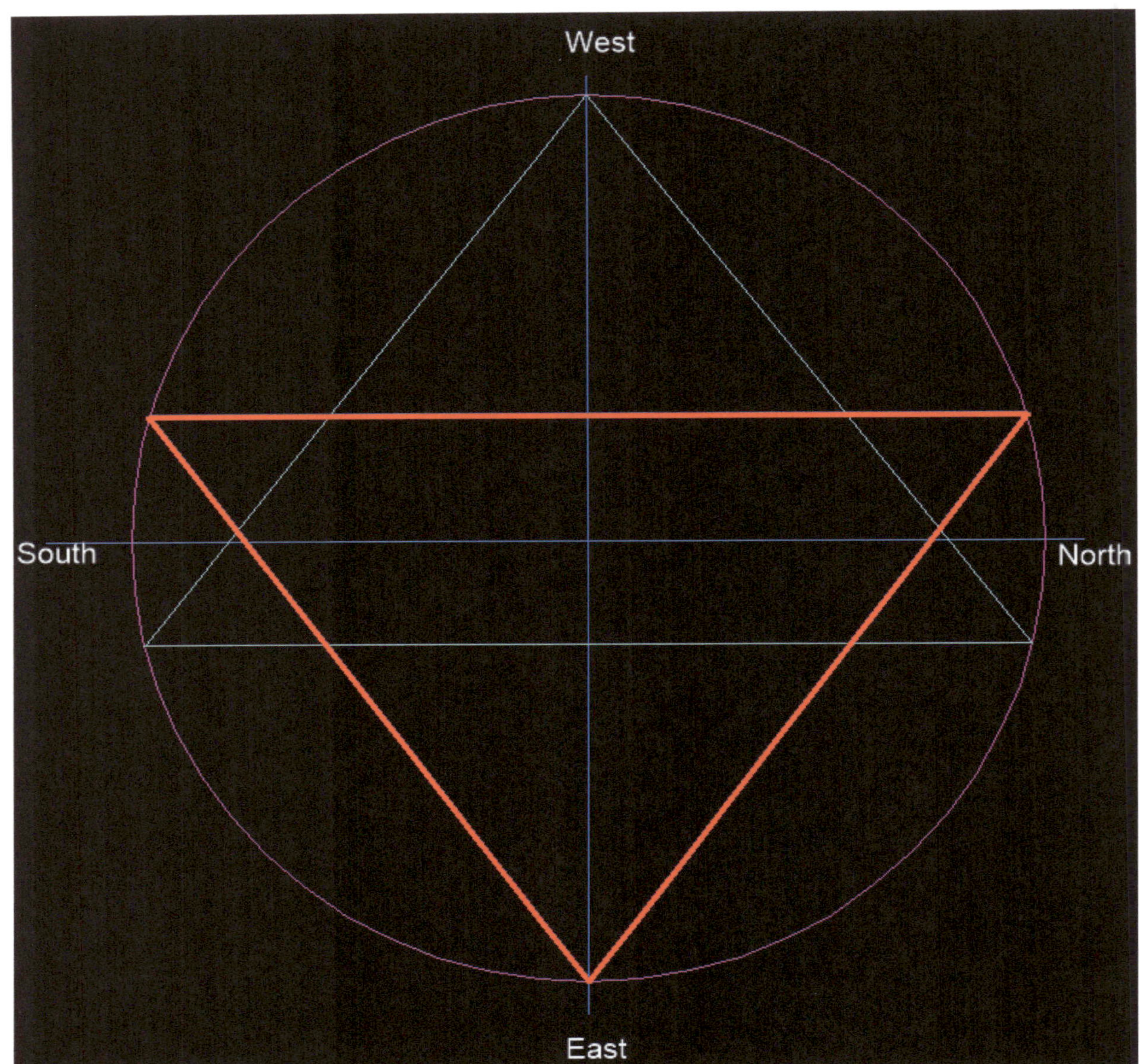

Figure 20 **T1d Completed**

17 Dimensions Largest Downward Apex Triangle T1d

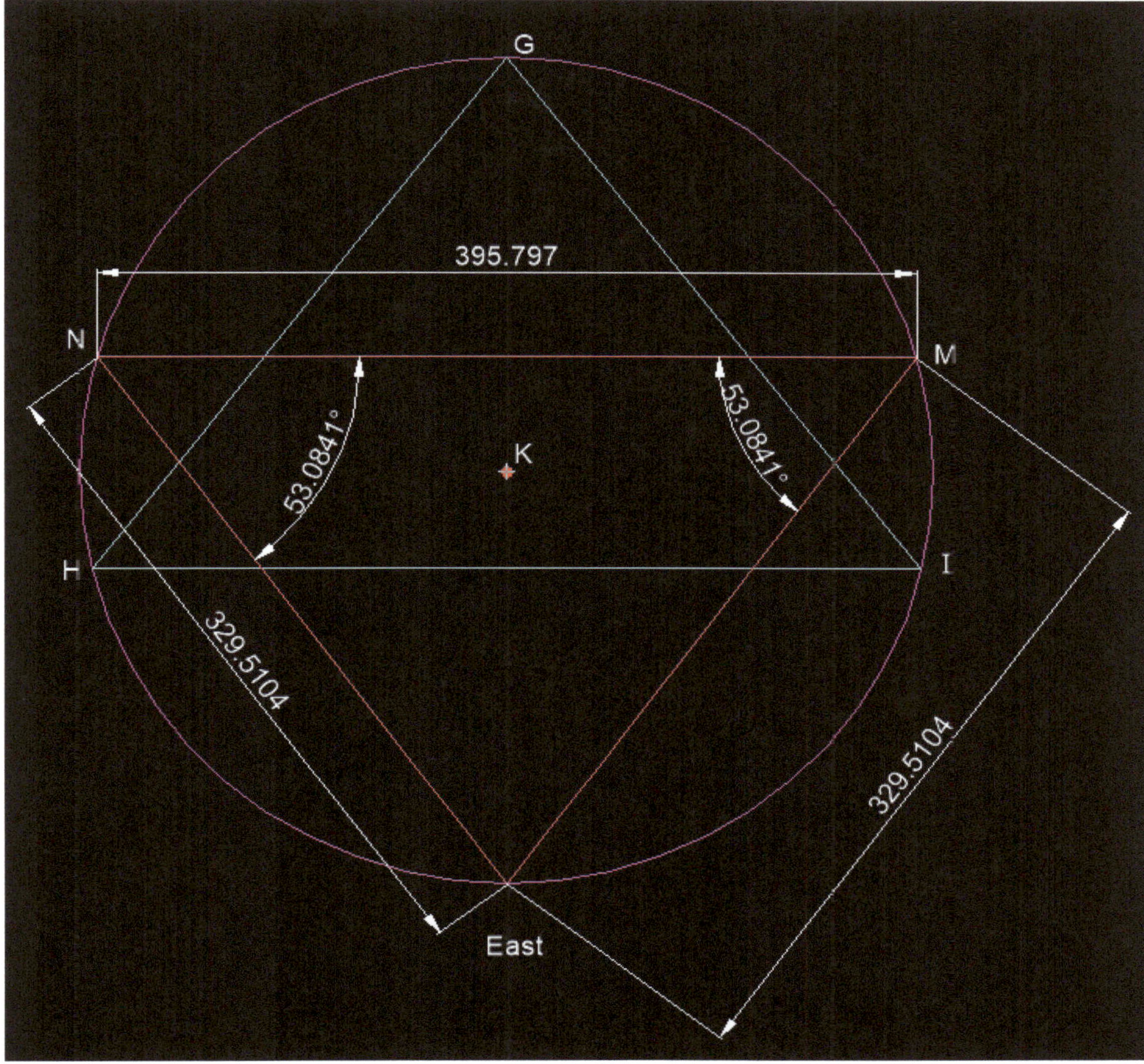

Figure 21 **T1d Dimensions**

Notes
1. T1d is slightly larger than T1u in area. These triangles are NOT Mirror copies.
2. T1d does not have precise Golden Ratio. Only T1u has the precise Golden Ratio.
3. T1d dimensions, Isosceles sides = 329.5104 each, Isosceles Angles = 53.0841° each, base = 395.797 mm. Area = 52139.1334 mm^2

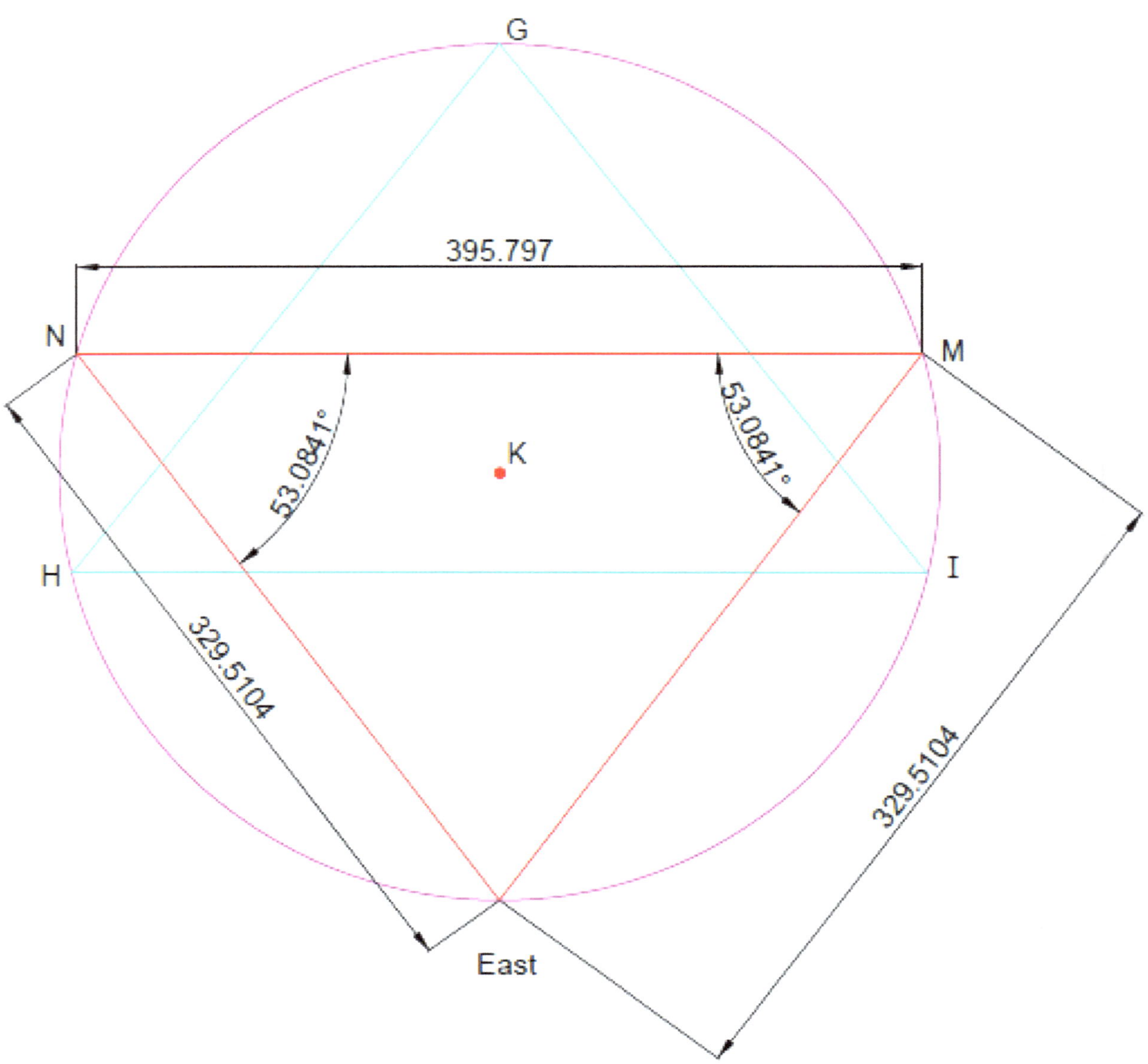

G
N
M
H
I
K
East
395.797
53.0841°
53.0841°
329.5104
329.5104

18 Cross Hair

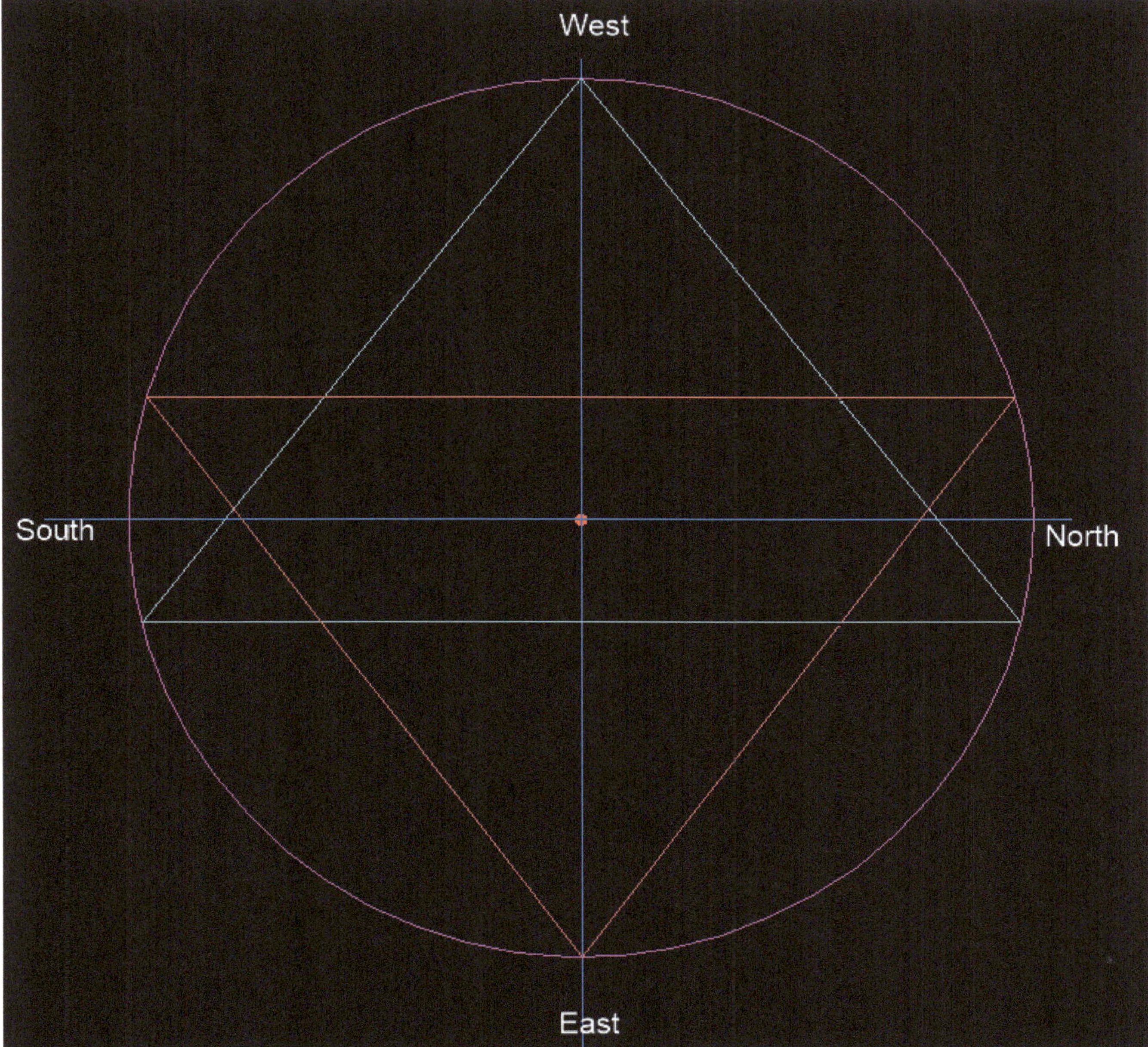

Figure 22 **Cross Hair Basic**

1. Draw the cross hair Vertical Horizontal by using the **Center Mark** of circle.
2. Notice that the cross hair passes through the center of circle, and it does **not** pass through the intersection of T1u and T1d. The intersection of T1u and T1d is **slightly above** the cross hair horizontal.
3. This is also to indicate that T1u and T1d are **not** mirror copy.
4. The cross hair intersection is the exact center of the circle, but it is **not** the Bindu.

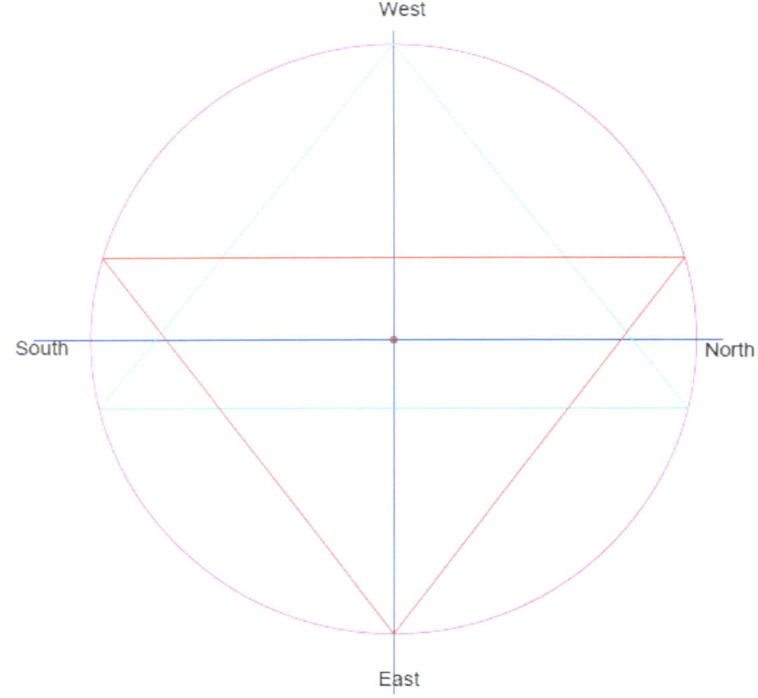

West
South
North
East

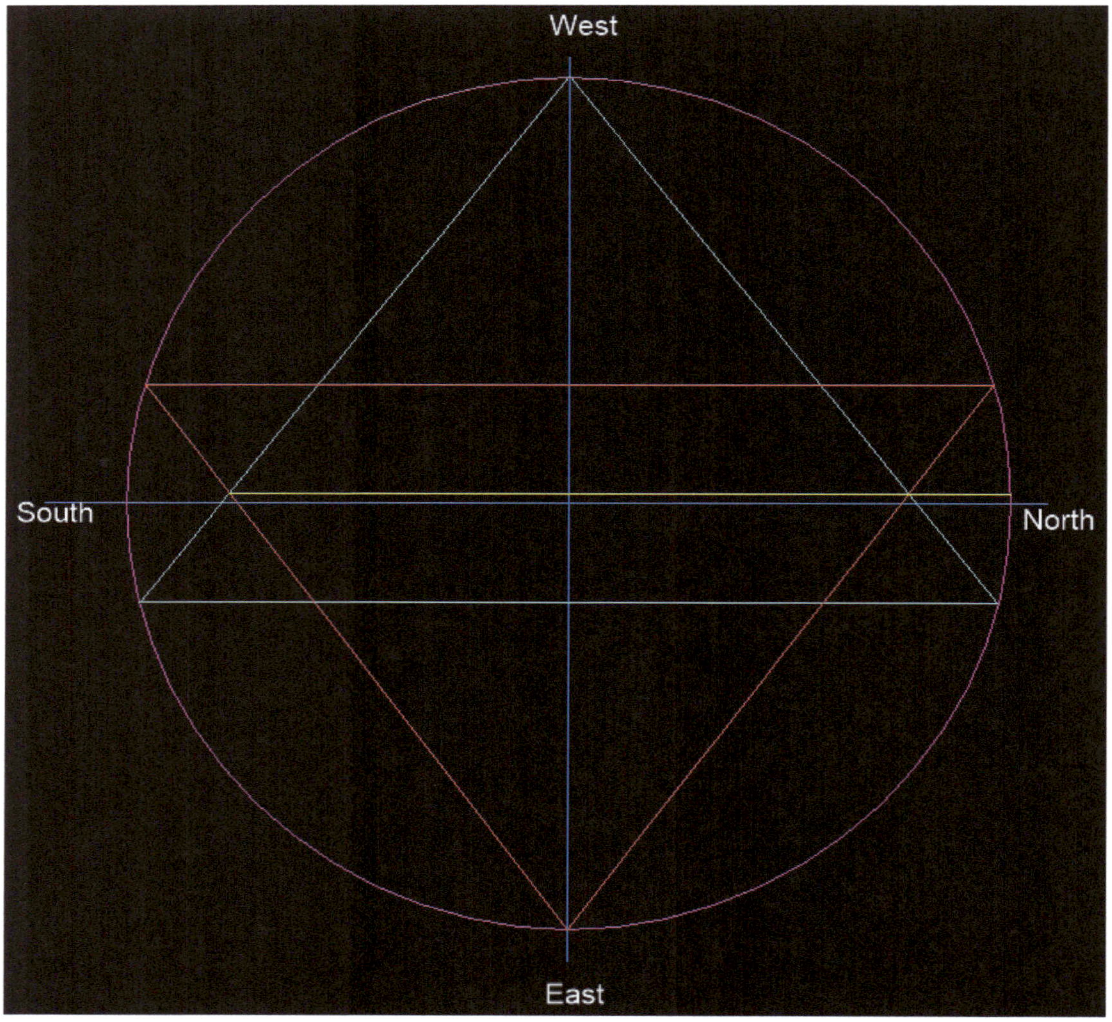

West
South
North
East

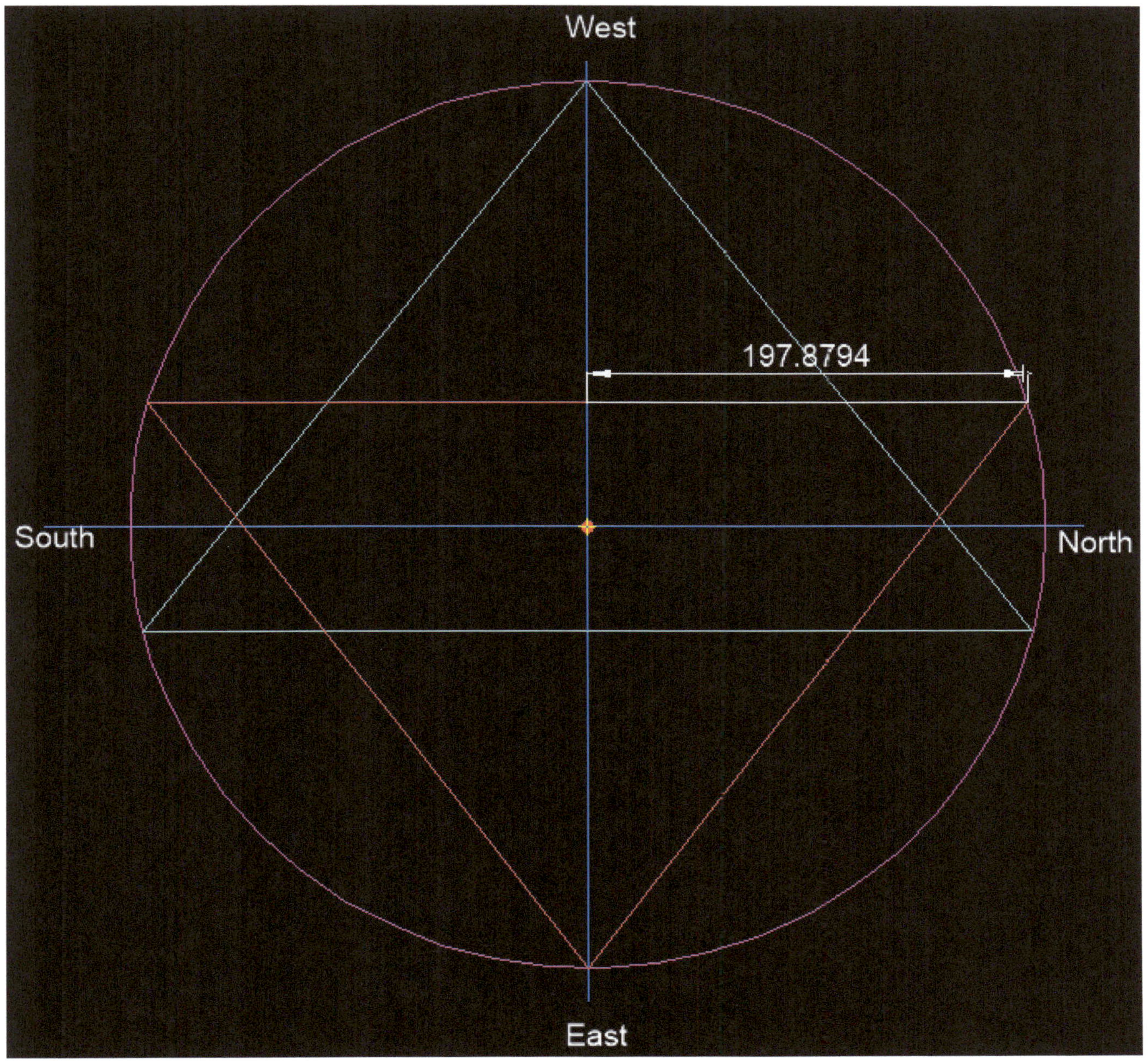

West
South
North
East
197.8794

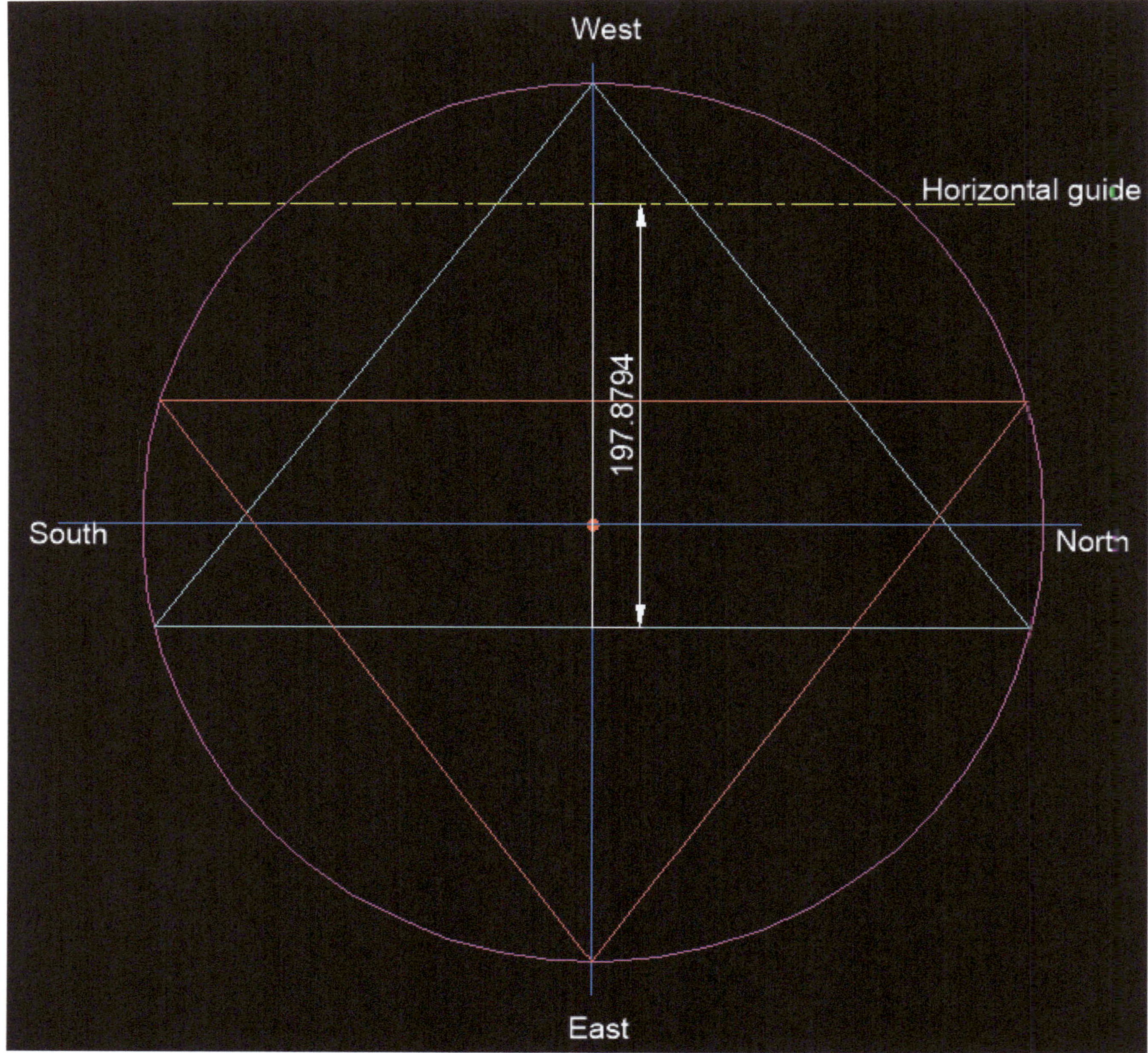

Figure 23 **Finding Apex of T2u**

1. Mark the **half** base line length of T1d.
2. Then **flip** it vertical and **align** its lower tip with intersection of T1u with Horizontal cross hair.
3. This gives the mark for the **Apex** of **new** Upward Apex Triangle **T2u**.
(4. The horizontal guide shall also be needed for drawing the base line of Triangle T3d later.)

20 Guides for sides of T2u

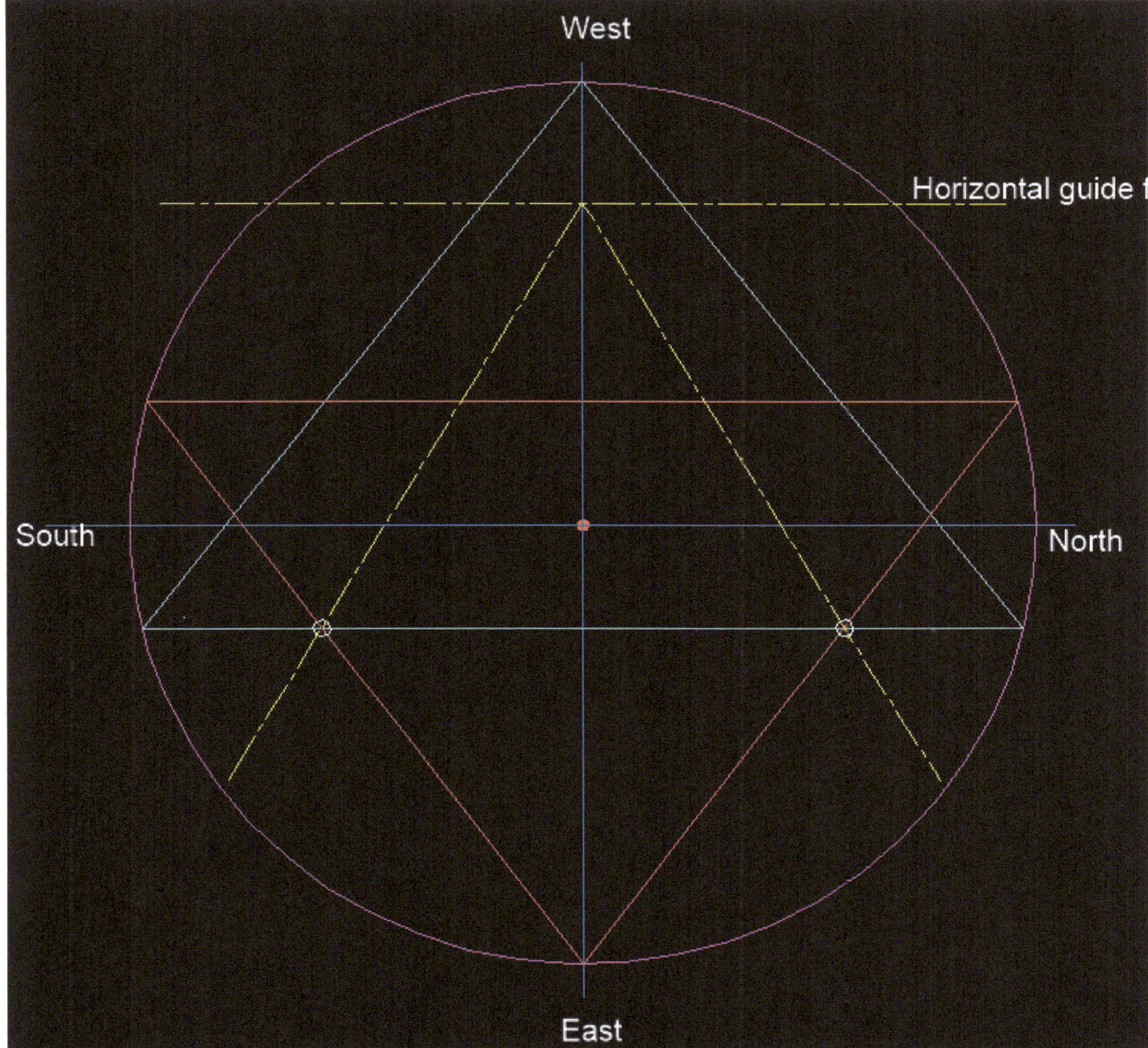

Figure 24 **Guides for T2u sides**

1. Mark the **half** base line length of red Triangle T1d.
2. Then **flip** it vertical and **align** its lower tip with intersection of T1u with Horizontal cross hair.
3. This gives the mark for the **Apex** of **new** Upward Apex Triangle **T2u**.
(4. The horizontal guide shall also be needed for drawing the base line of Triangle T3d later.)
5. Using this Apex and the lower sandhi Points of T1u, T1d draw the guides for the sides of Triangle T2u.

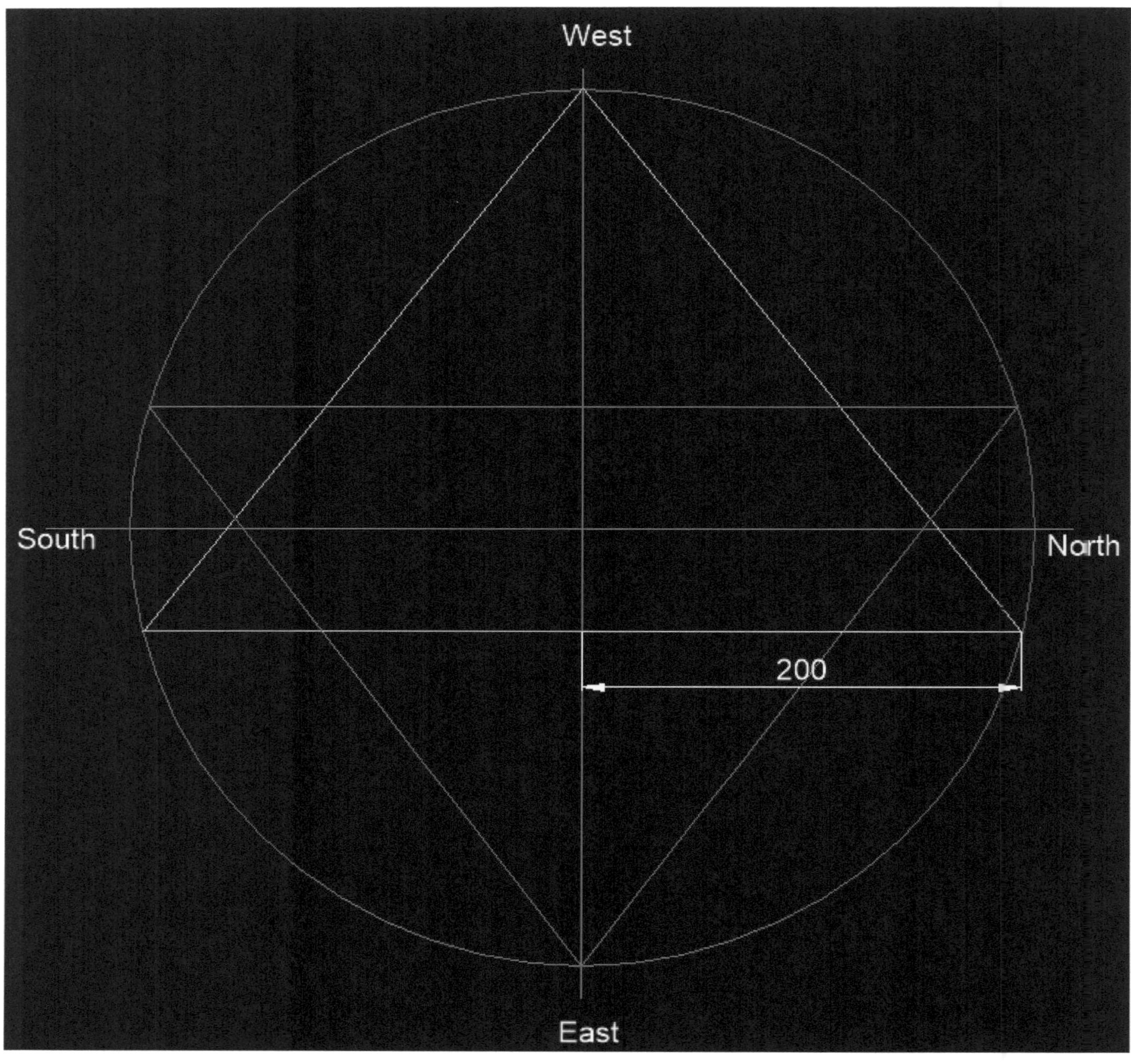

West
South
North
East
200

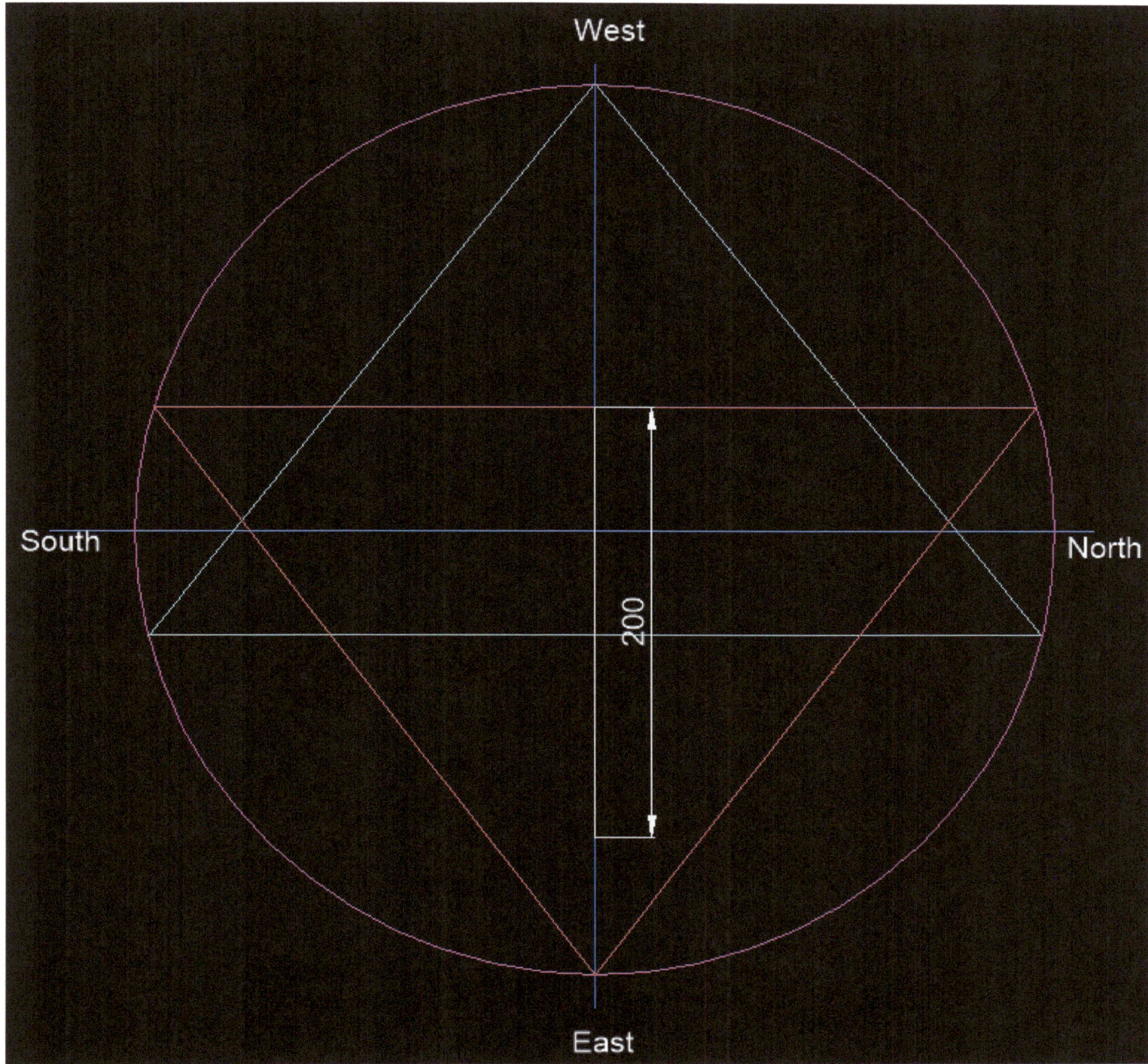

Figure 25 **Finding Apex for T2d**

1. Mark the **half** base line length of T1u.
2. Then **flip** it vertical and **align** its lower tip with intersection of T1d with Horizontal cross hair.
3. This gives the mark for the **Apex** of **new** Upward Apex Triangle **T2d**.
(4. The horizontal guide shall also be needed for drawing the base line of Triangle T3u later.)

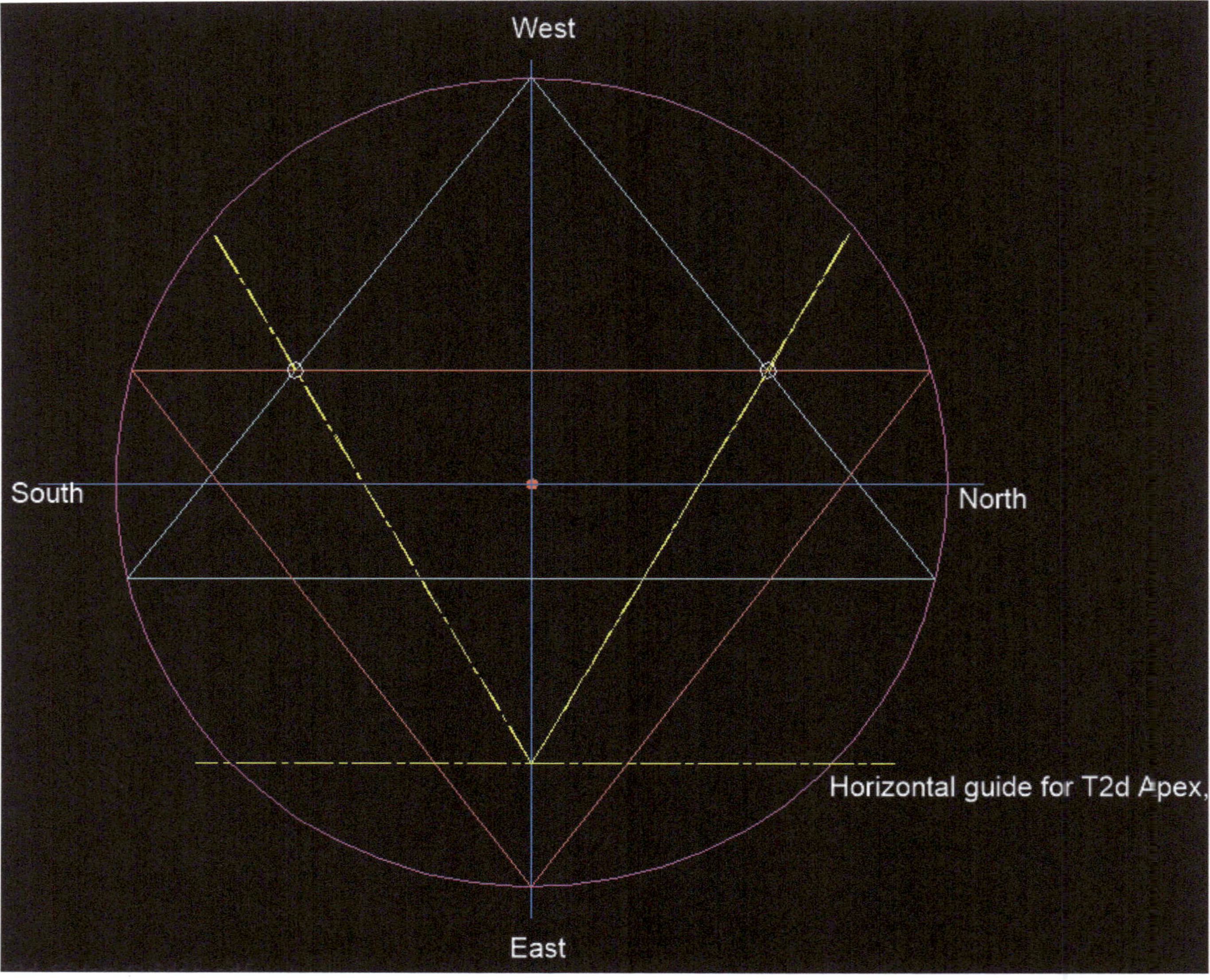

Figure 26 **3 4 5 Guides for T2d sides**

1. Mark the **half** base line length of red Triangle T1d.
2. Then **flip** it vertical and **align** its lower tip with intersection of T1u with Horizontal cross hair.
3. This gives the mark for the **Apex** of **new** Upward Apex Triangle **T2u**.
(4. The horizontal guide shall also be needed for drawing the base line of Triangle T3u later.)
5. Using this Apex and the lower sandhi Points of T1u T1d, draw the guides for the sides of Triangle T2d.

As each new side gets drawn, we get the sandhi junction points for seeing the Apex or Sides for the next Triangle. Be aware of this.

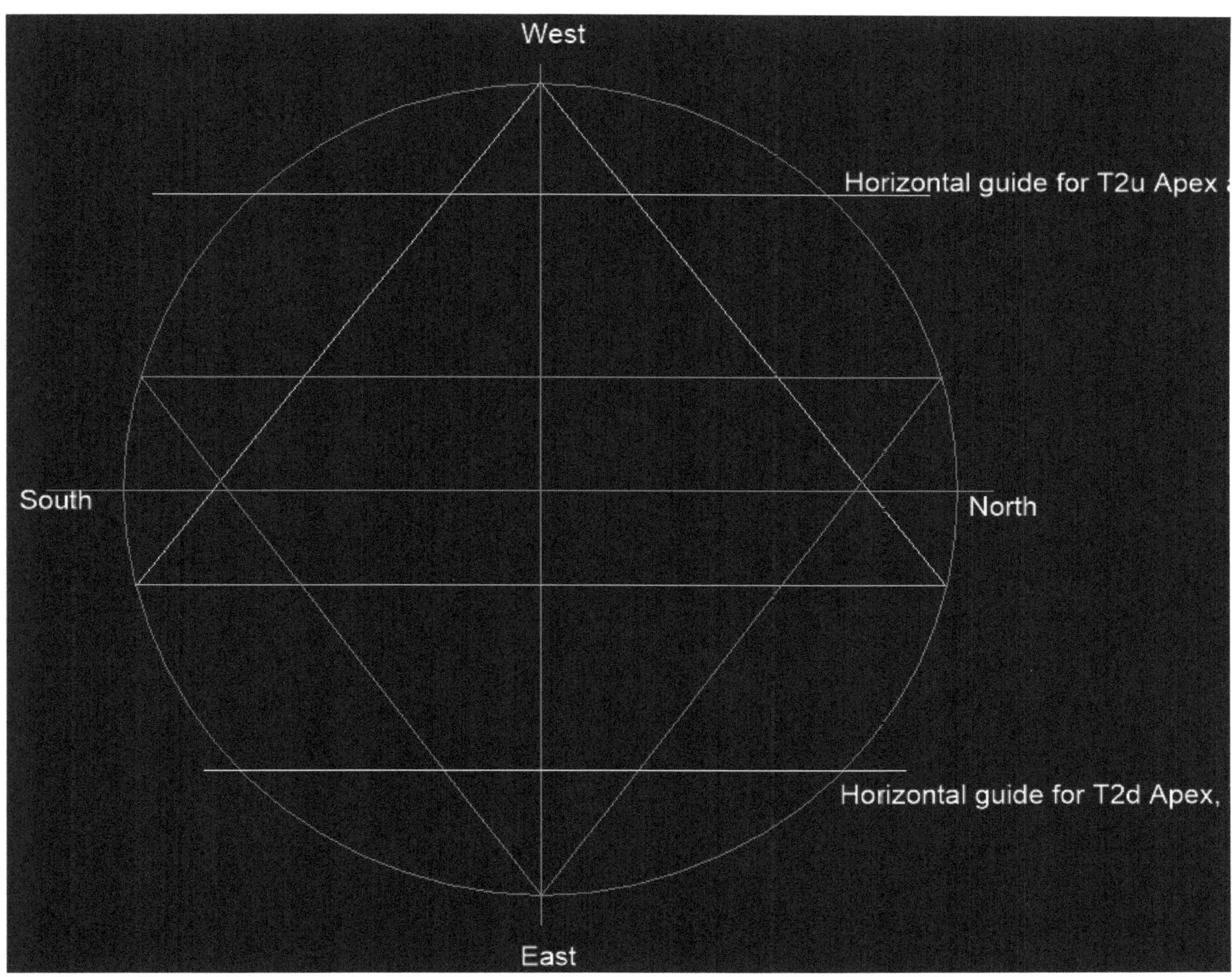

1. The top horizontal guide is for T2u Apex and T4d base.
2. The bottom horizontal guide is for T2d Apex and T4u base.

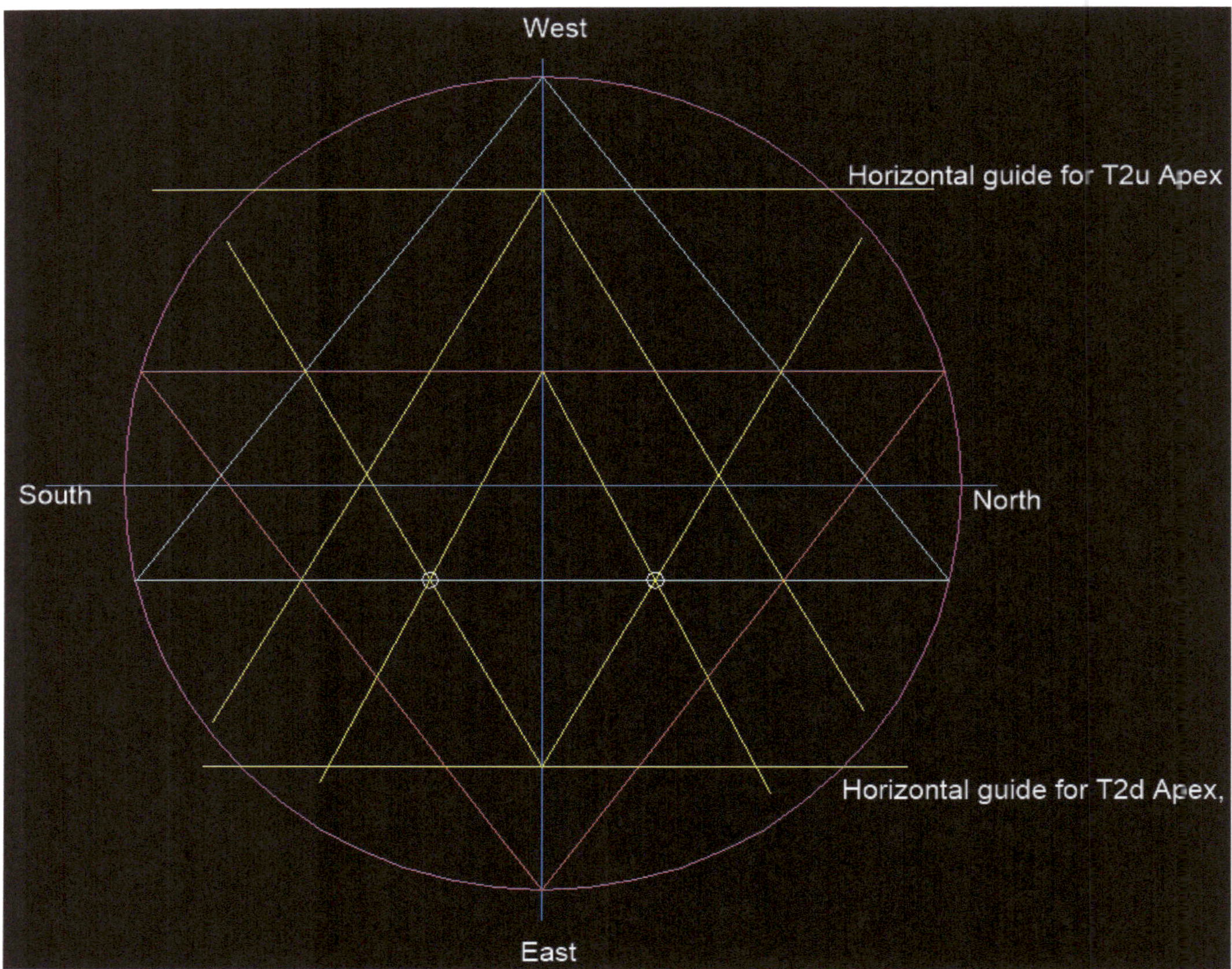

Figure 27 **Finding T4u**

24 Complete the seventh Triangle T4u

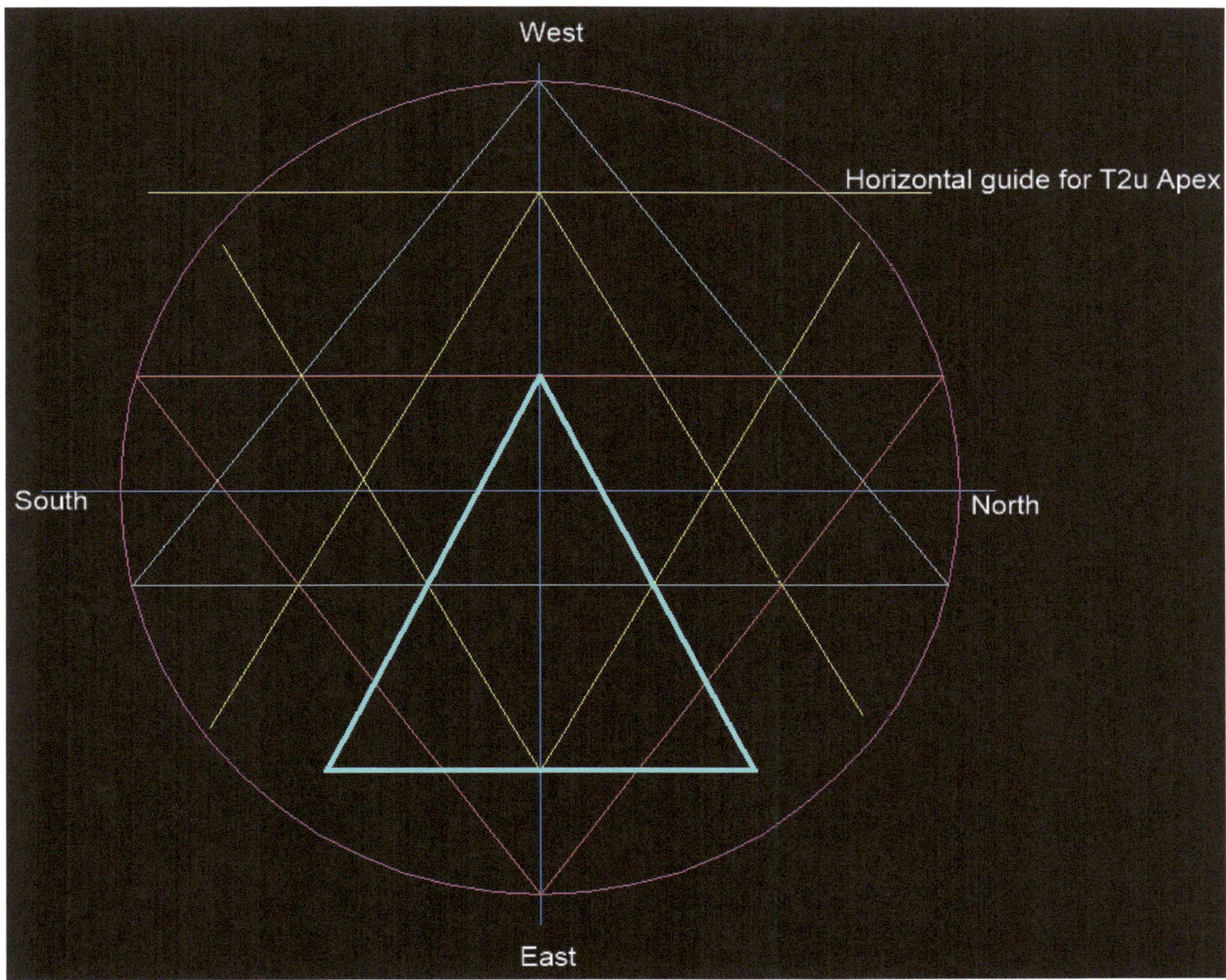

Figure 28 **T4u Completed**

25 Dimensions of seventh Triangle T4u

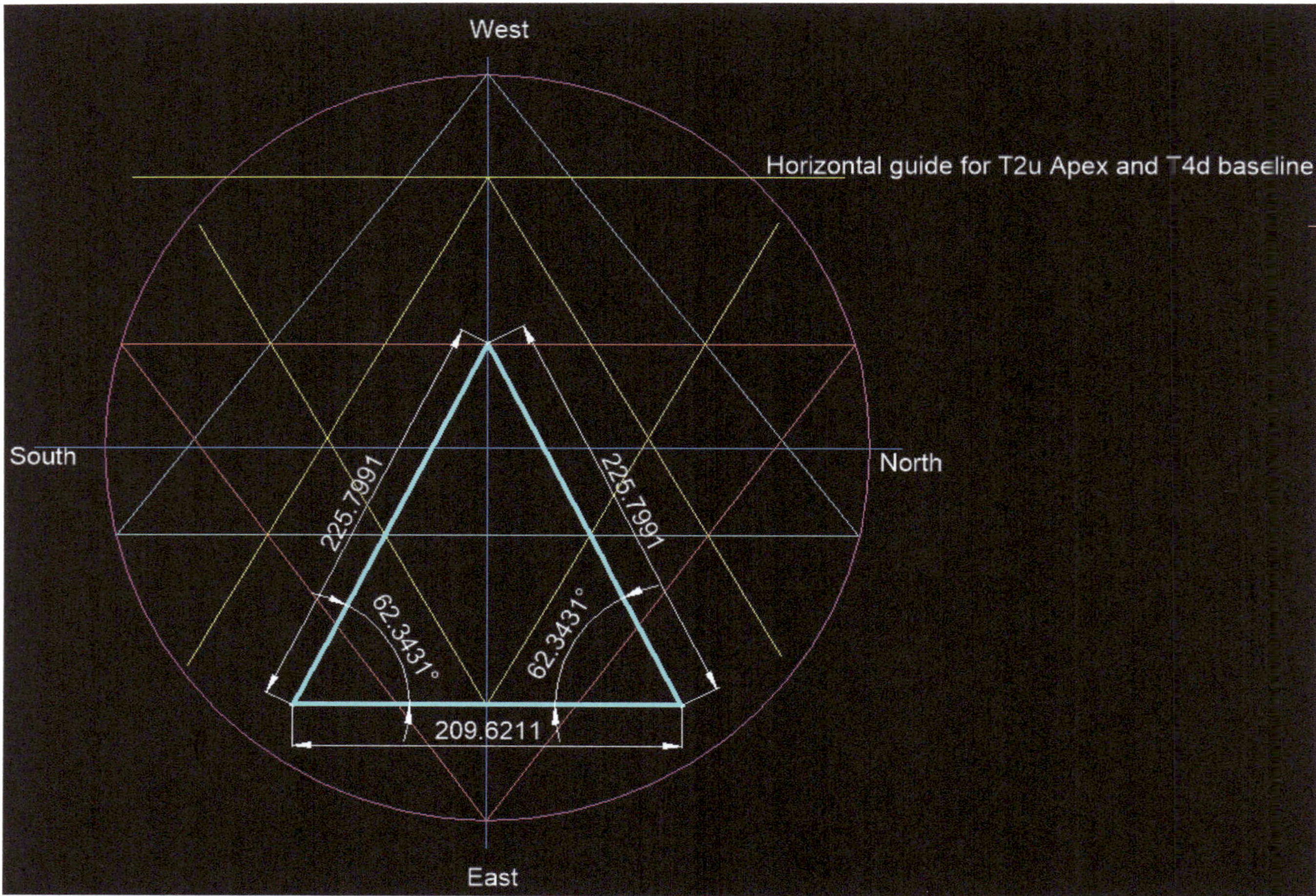

Figure 29 **T4u Dimensions**

T4u dimensions, Isosceles sides = 329.5104 each, Isosceles Angles = 53.0841° each, base = 395.797 mm. Area = 52139.13345 mm²

26 Base guide for third Triangle T2u

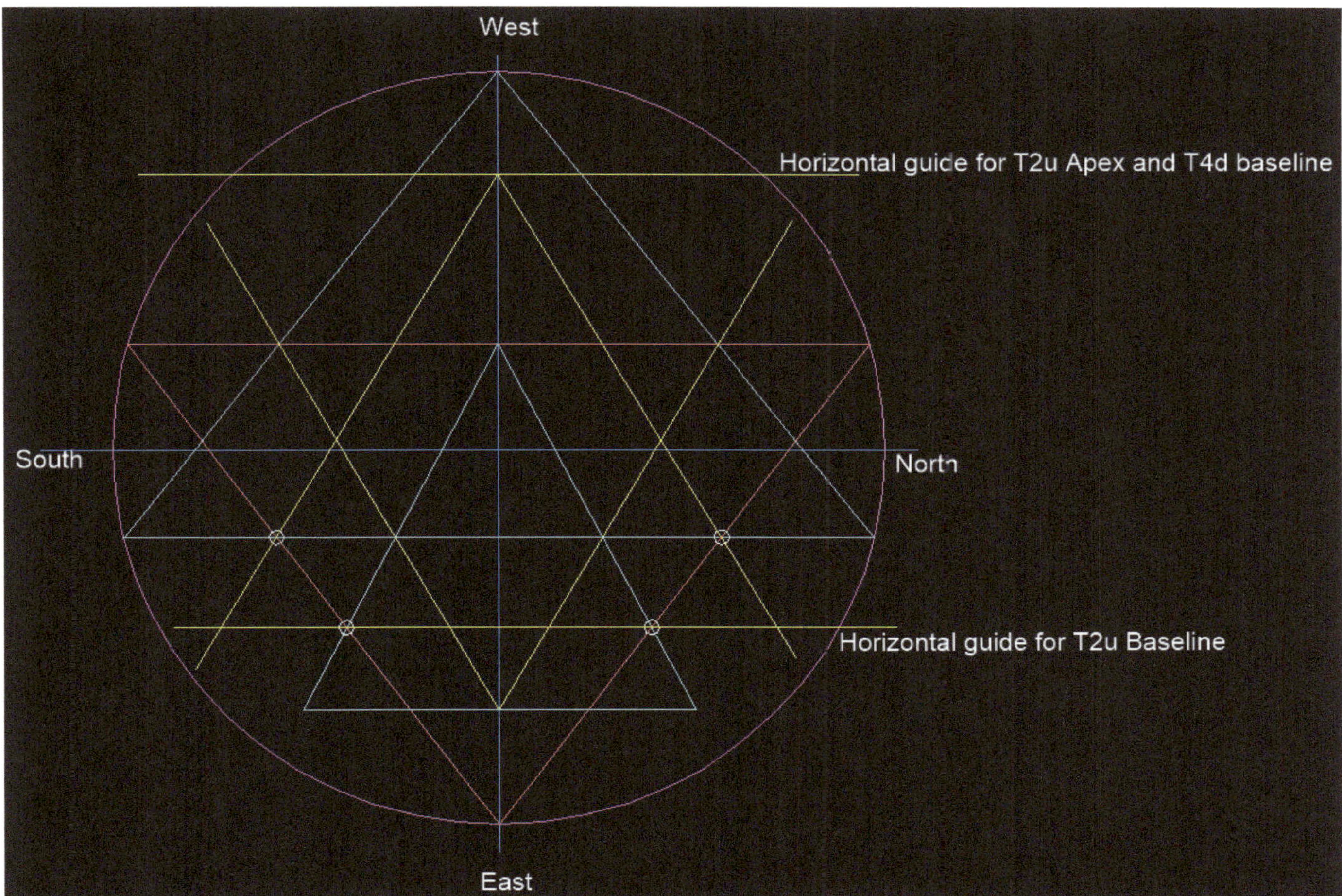

Figure 30 **Finding T2u**

27 Complete the third Triangle T2u

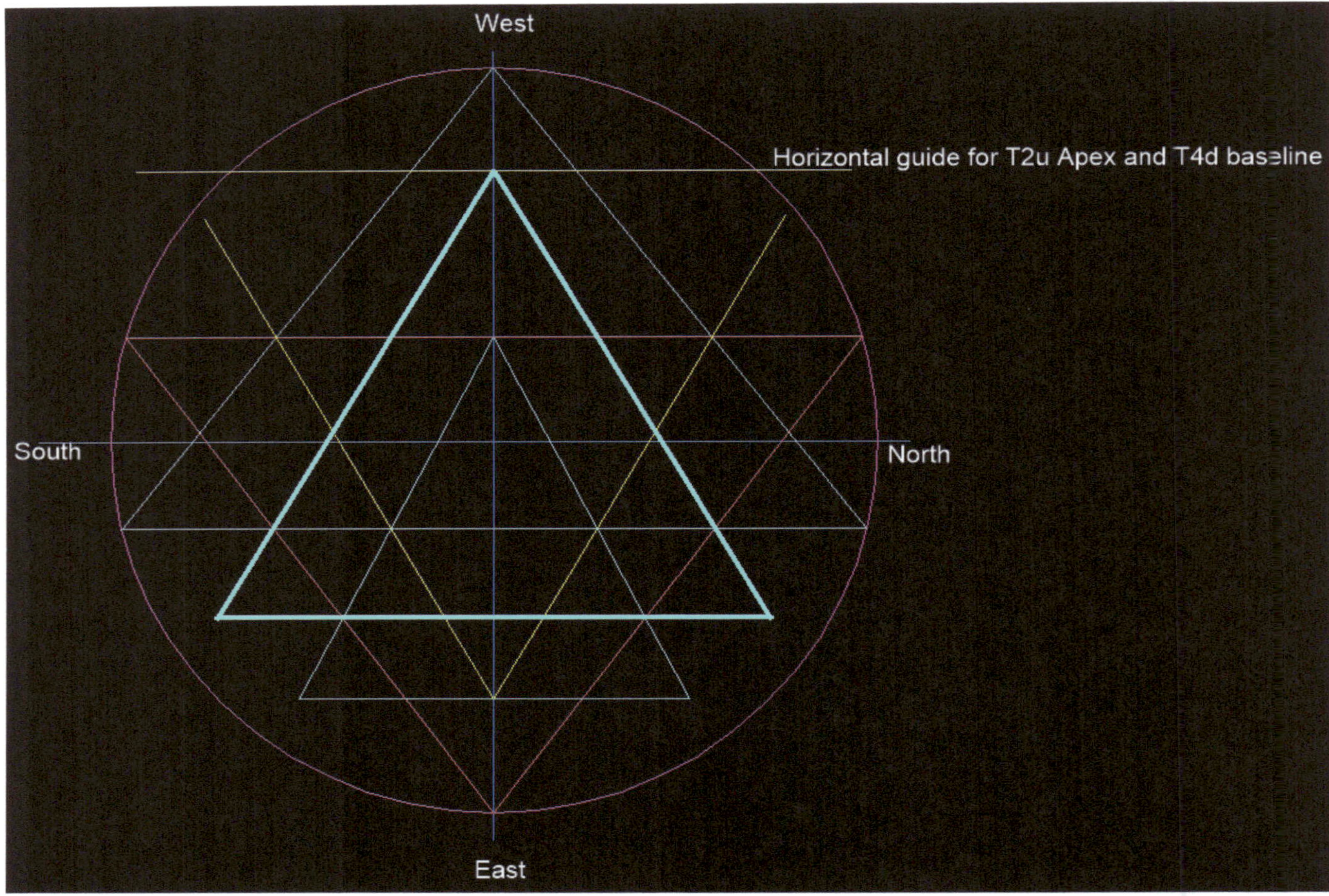

Figure 31 **T2u Completed**

28 Dimensions of third Triangle T2u

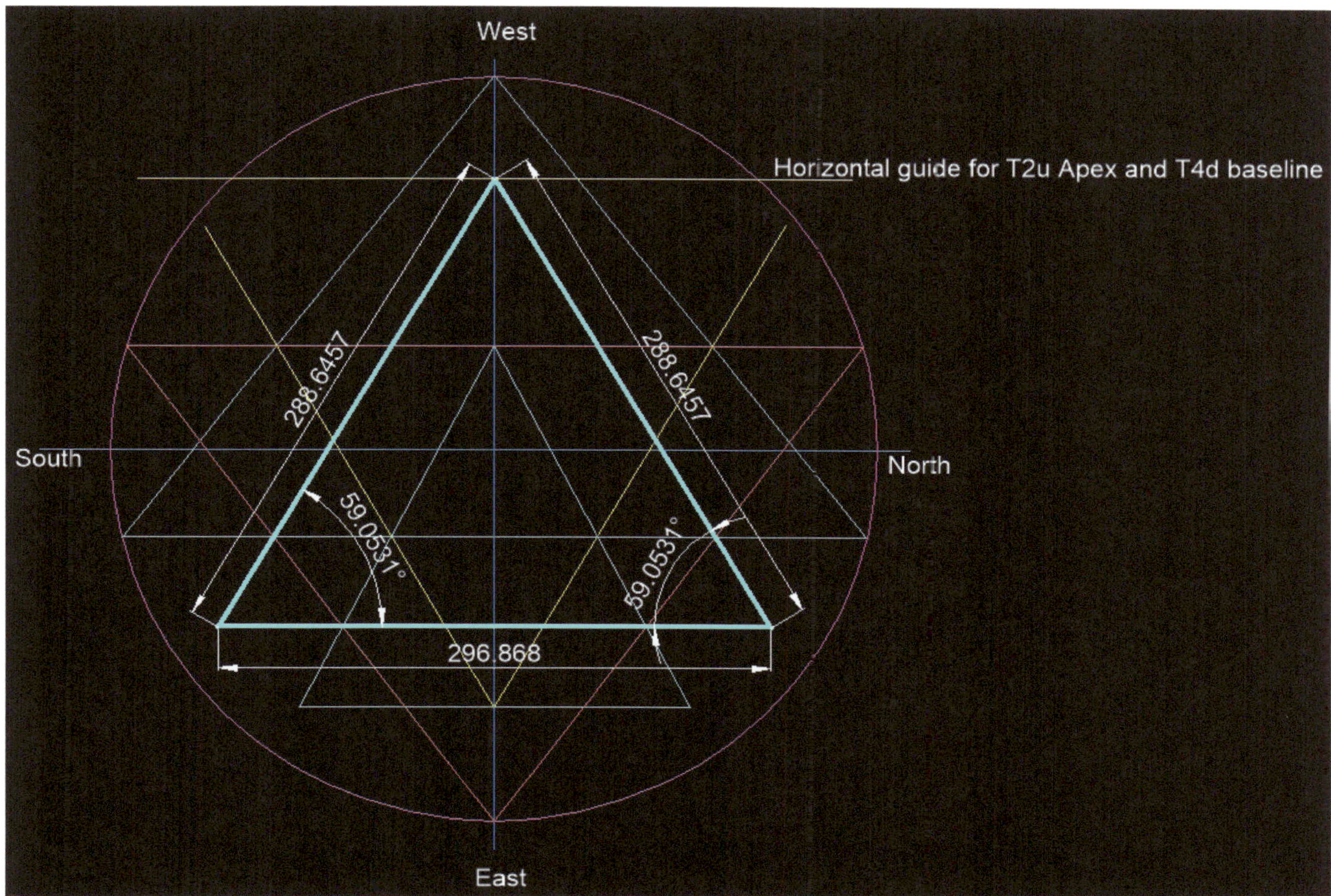

Figure 32 **T2u Dimensions**

T2u dimensions, Isosceles sides = 288.6457 each, Isosceles Angles = 59.0531° each, base = 296.868 mm. Area = 36745.64175 mm²

29 Finding the Apex of eighth Triangle T4d

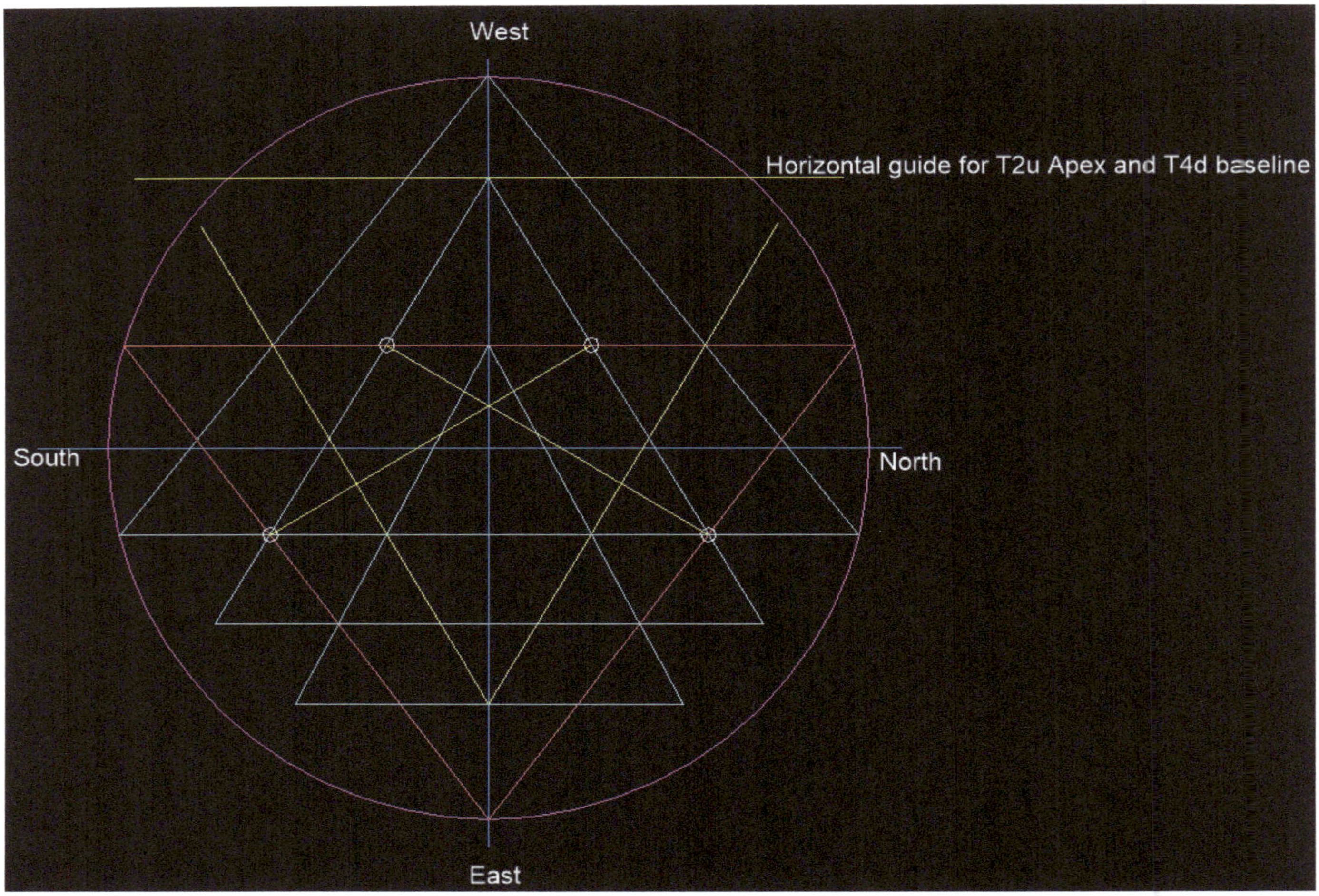

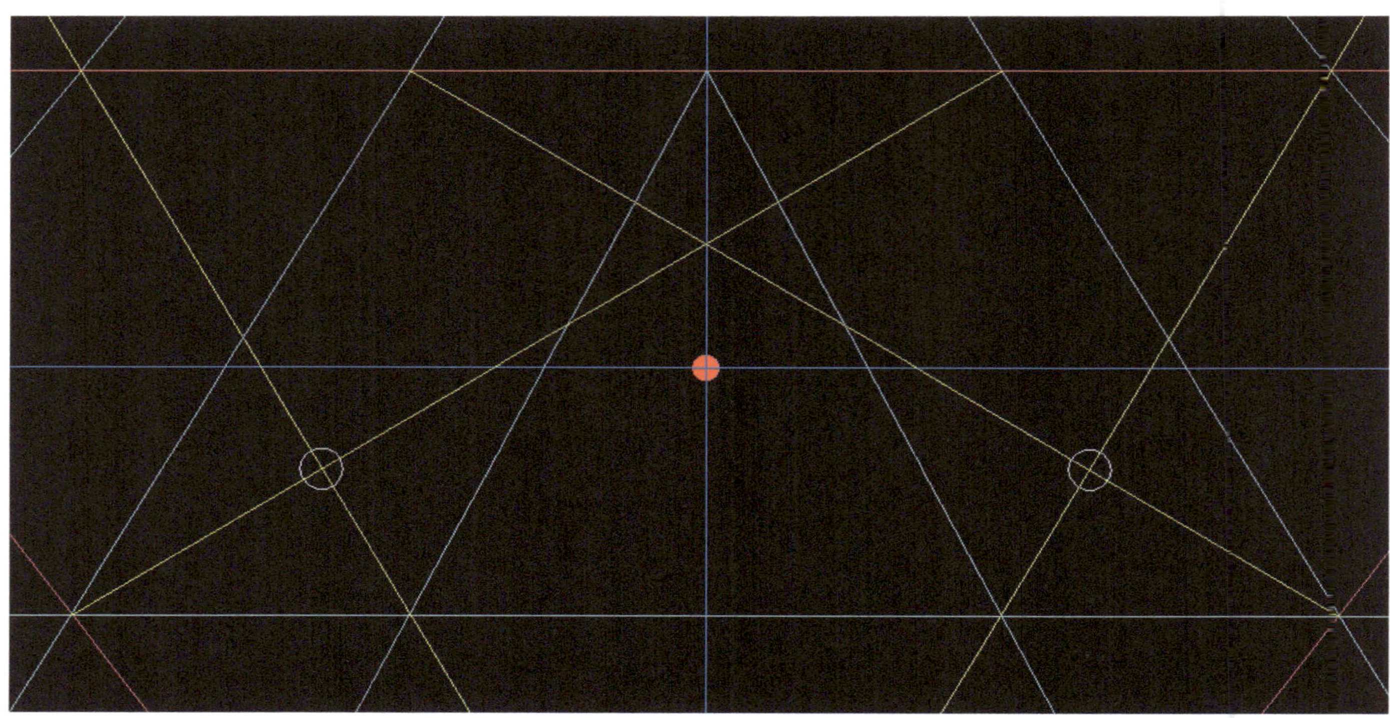

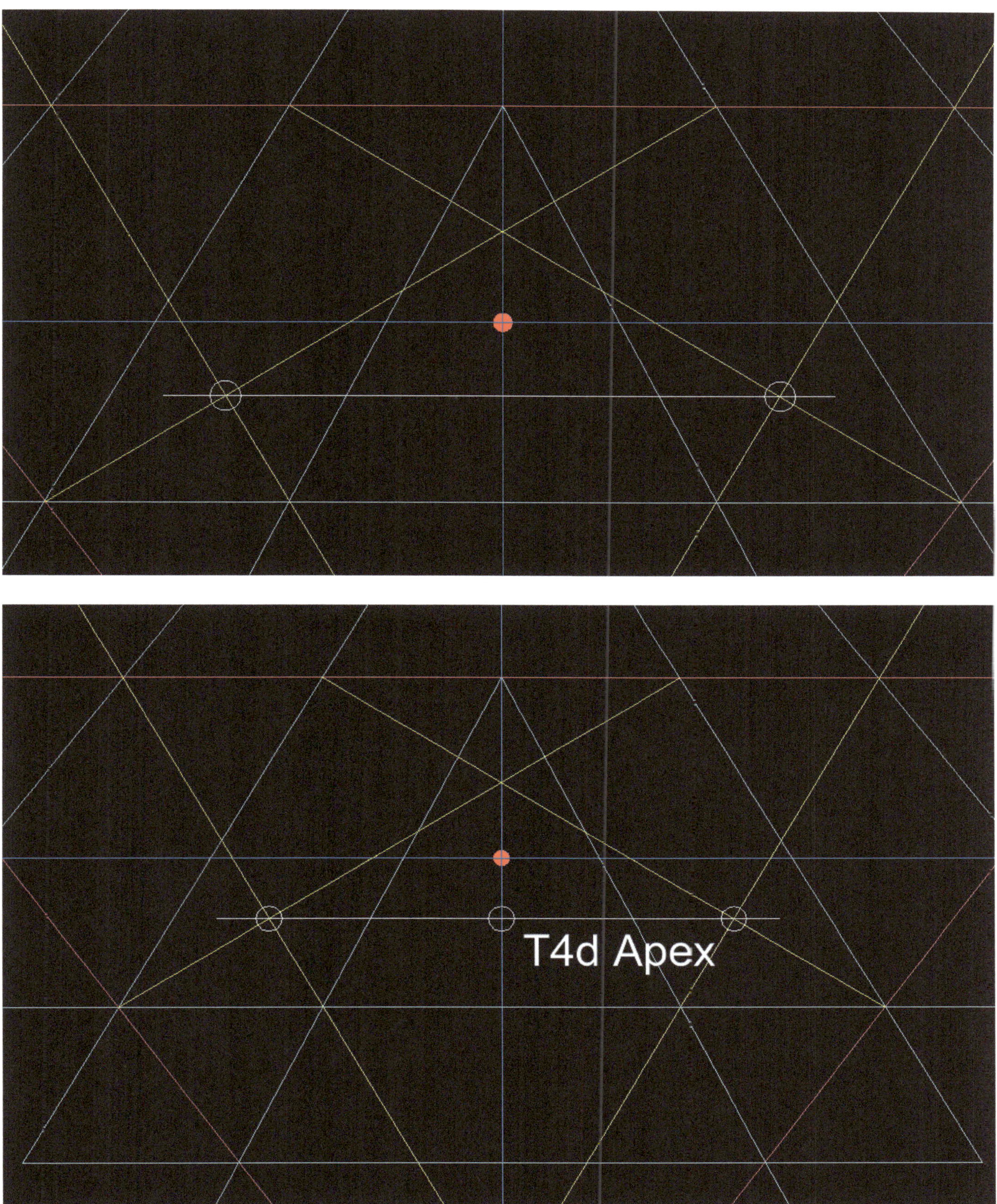

T4d Apex

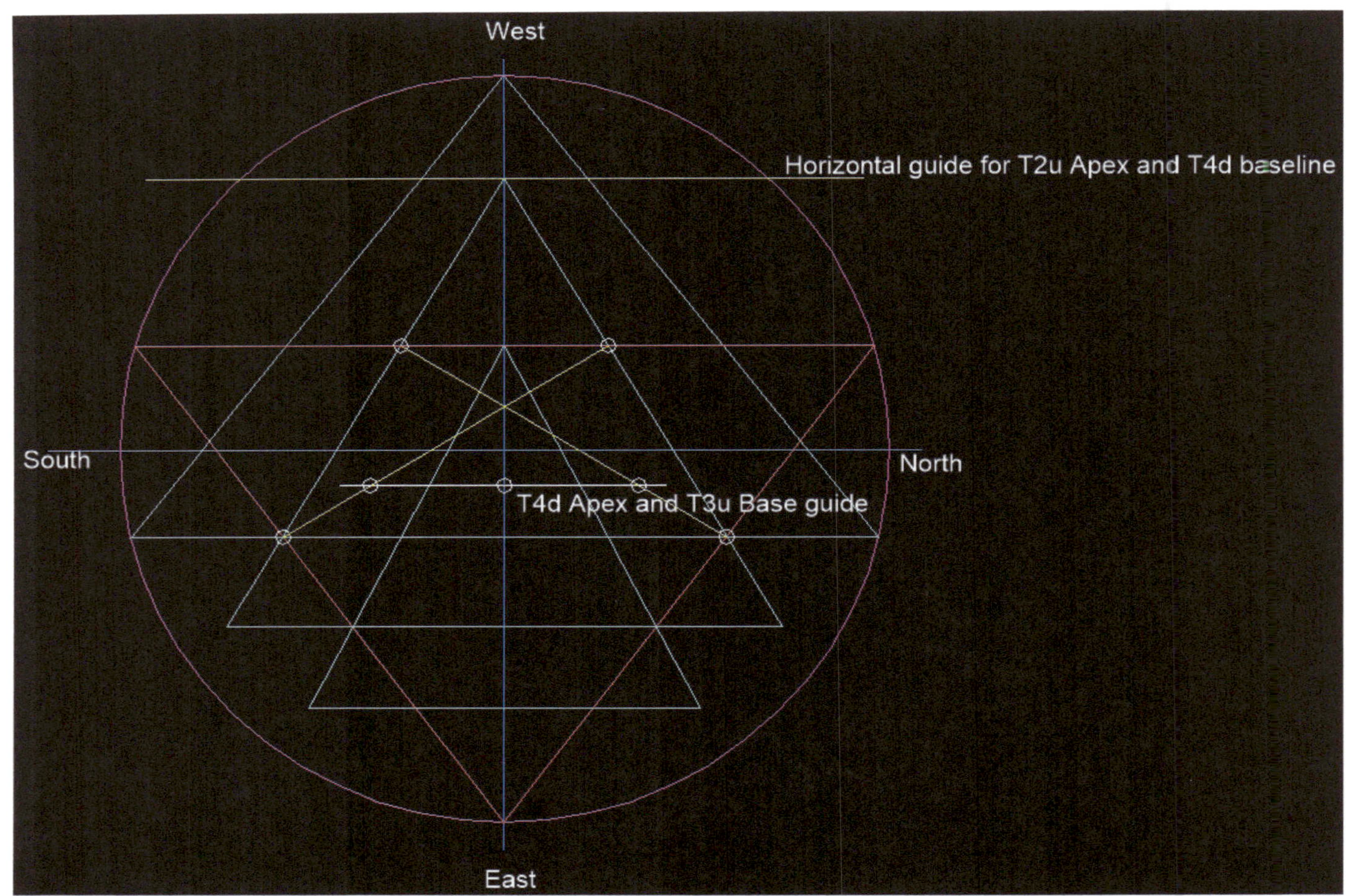

Figure 33 **Finding T4d Apex**

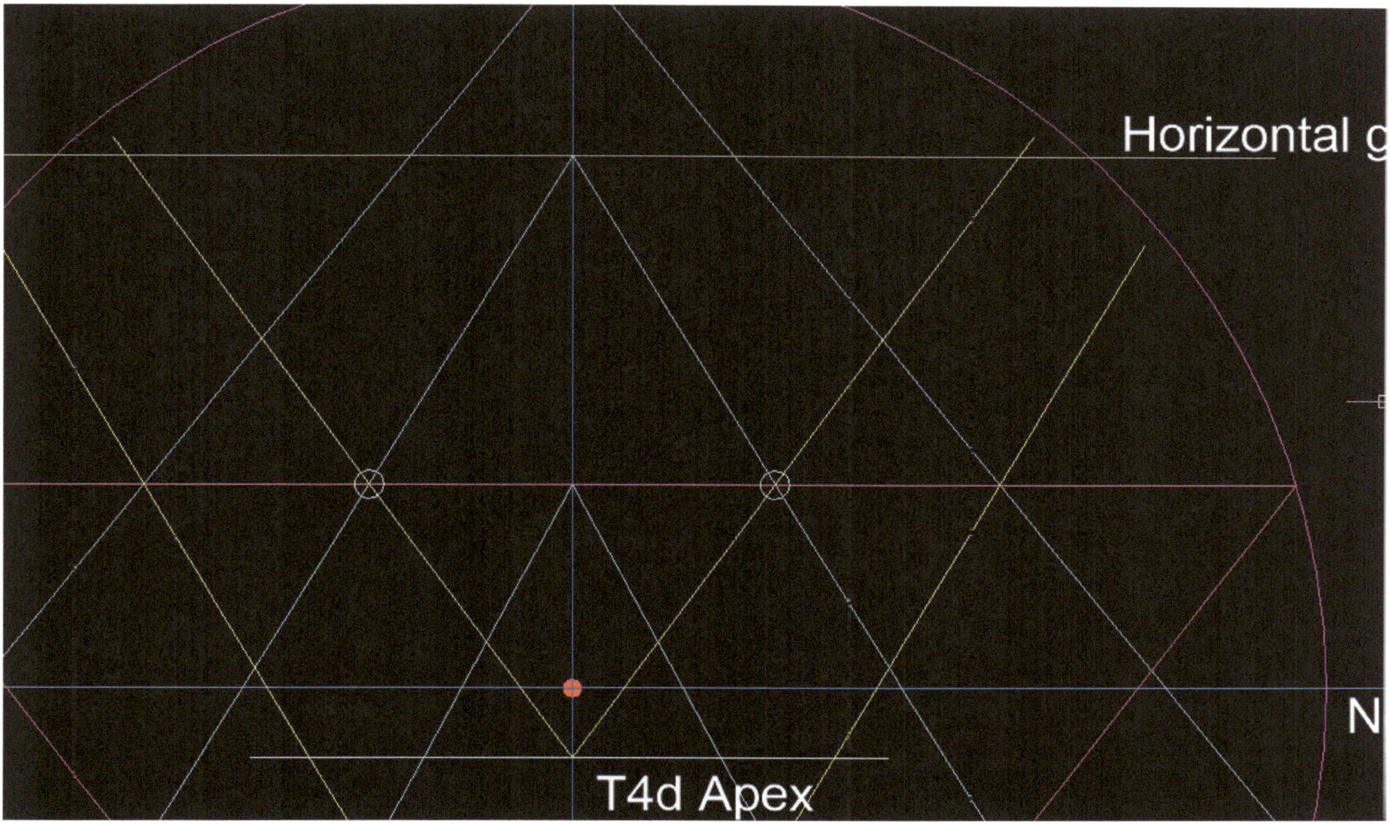

Figure 34 **Guides T4d sides**

31 Guides for base of eighth Triangle T4d

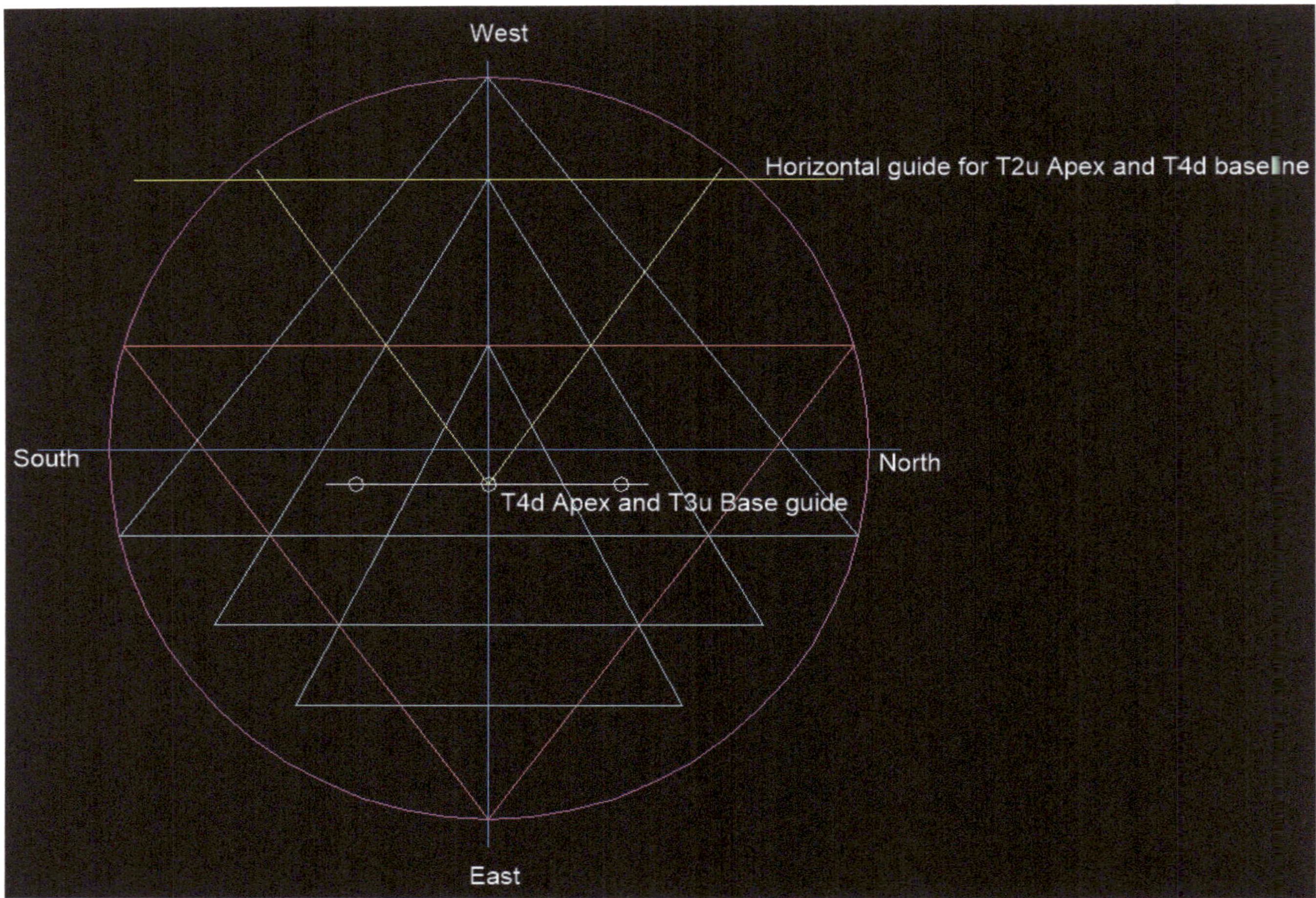

Figure 35 **Guide T4d Apex and baseline**

32 Complete the eighth Triangle T4d

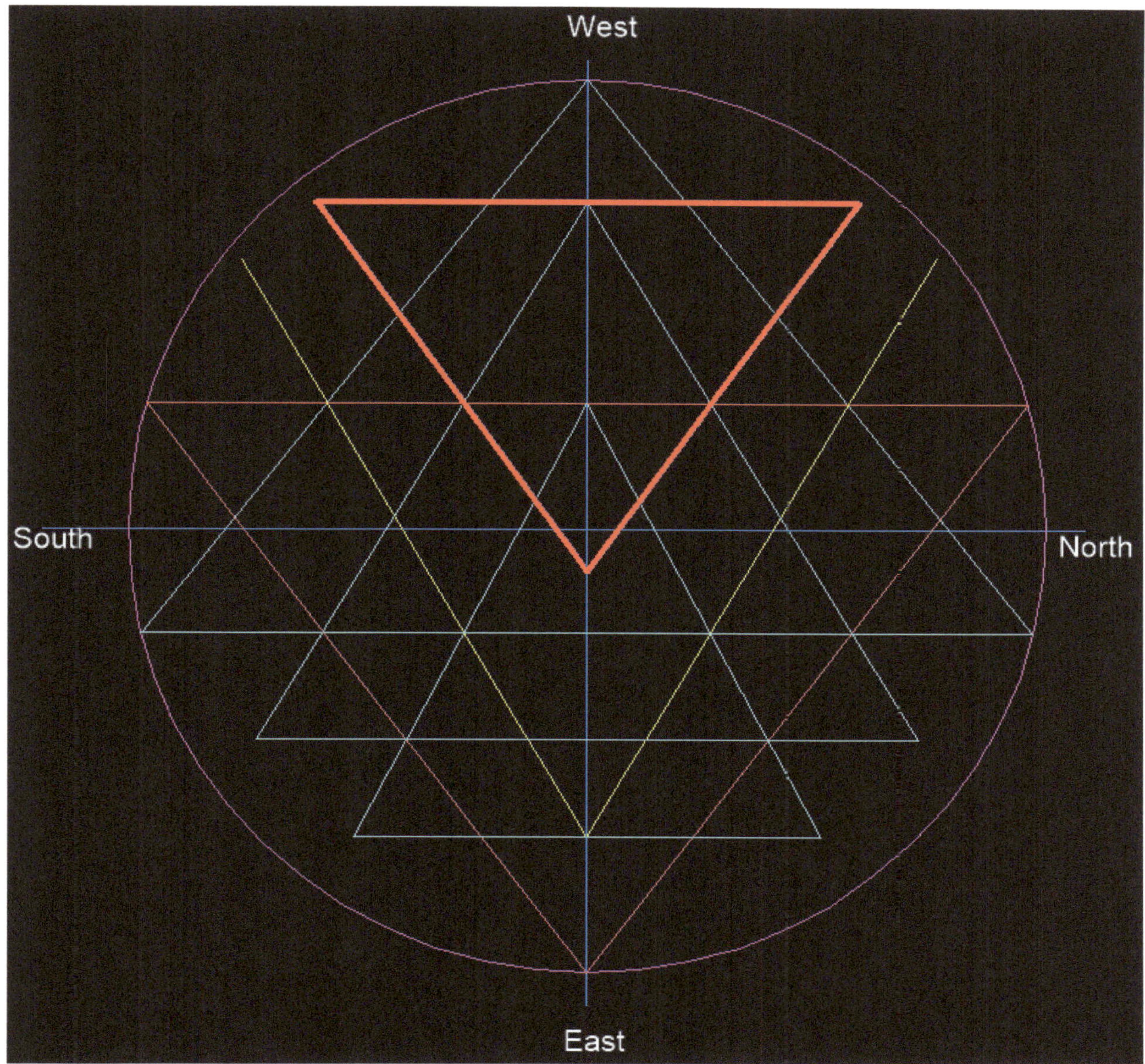

Figure 36 **T4d Completed**

33 Dimensions of eighth Triangle T4d

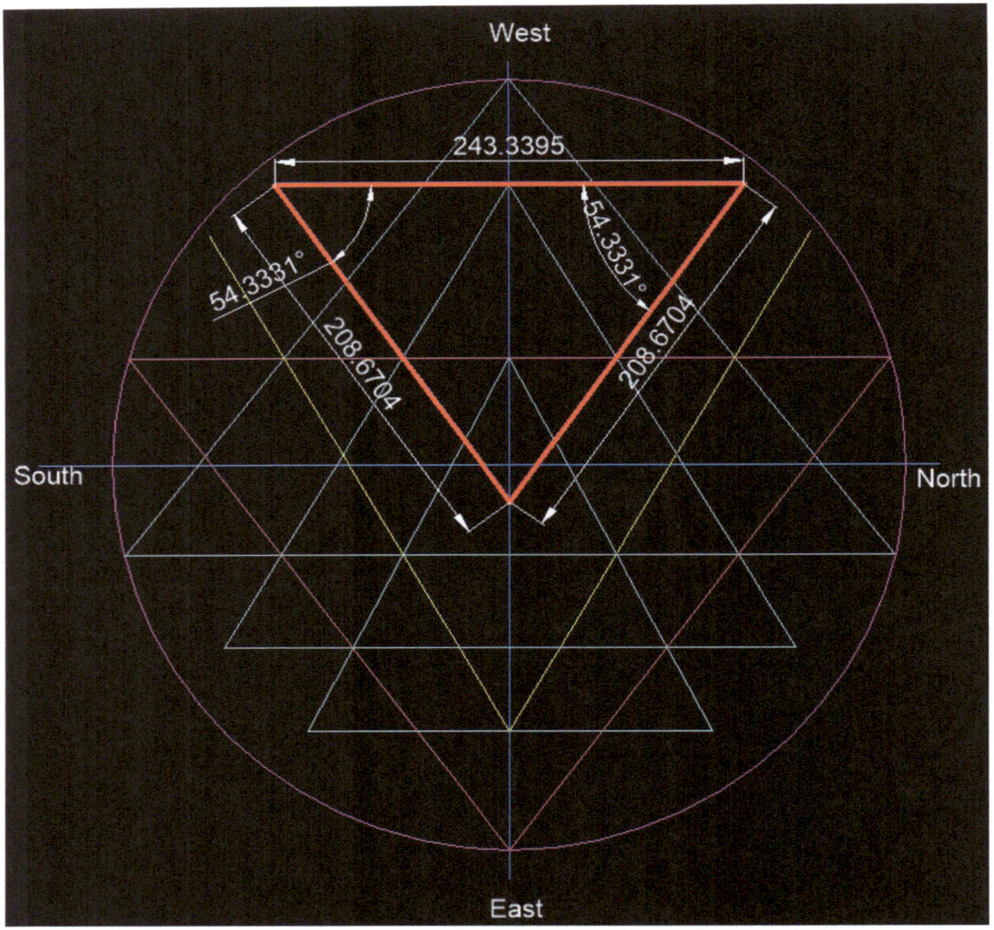

Figure 37 **T4d Dimensions**

T4d dimensions, Isosceles sides = 208.6704 each, Isosceles Angles = 54.3331° each, base = 243.3395 mm. Area = 20626.45269 mm²

34 Finding the base of fourth Triangle T2d

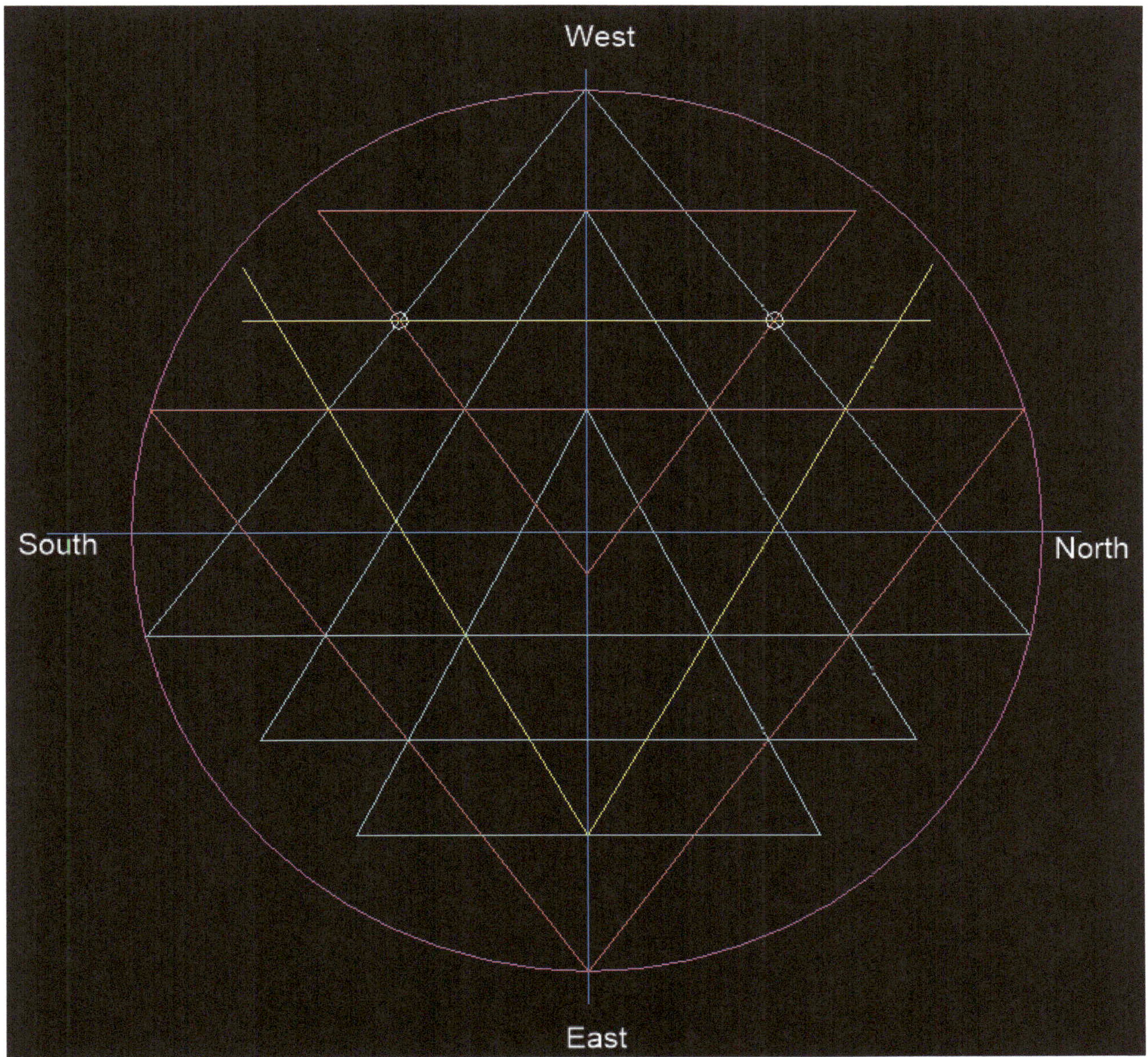

Figure 38 **Finding T2d**

35 Complete the fourth Triangle T2d

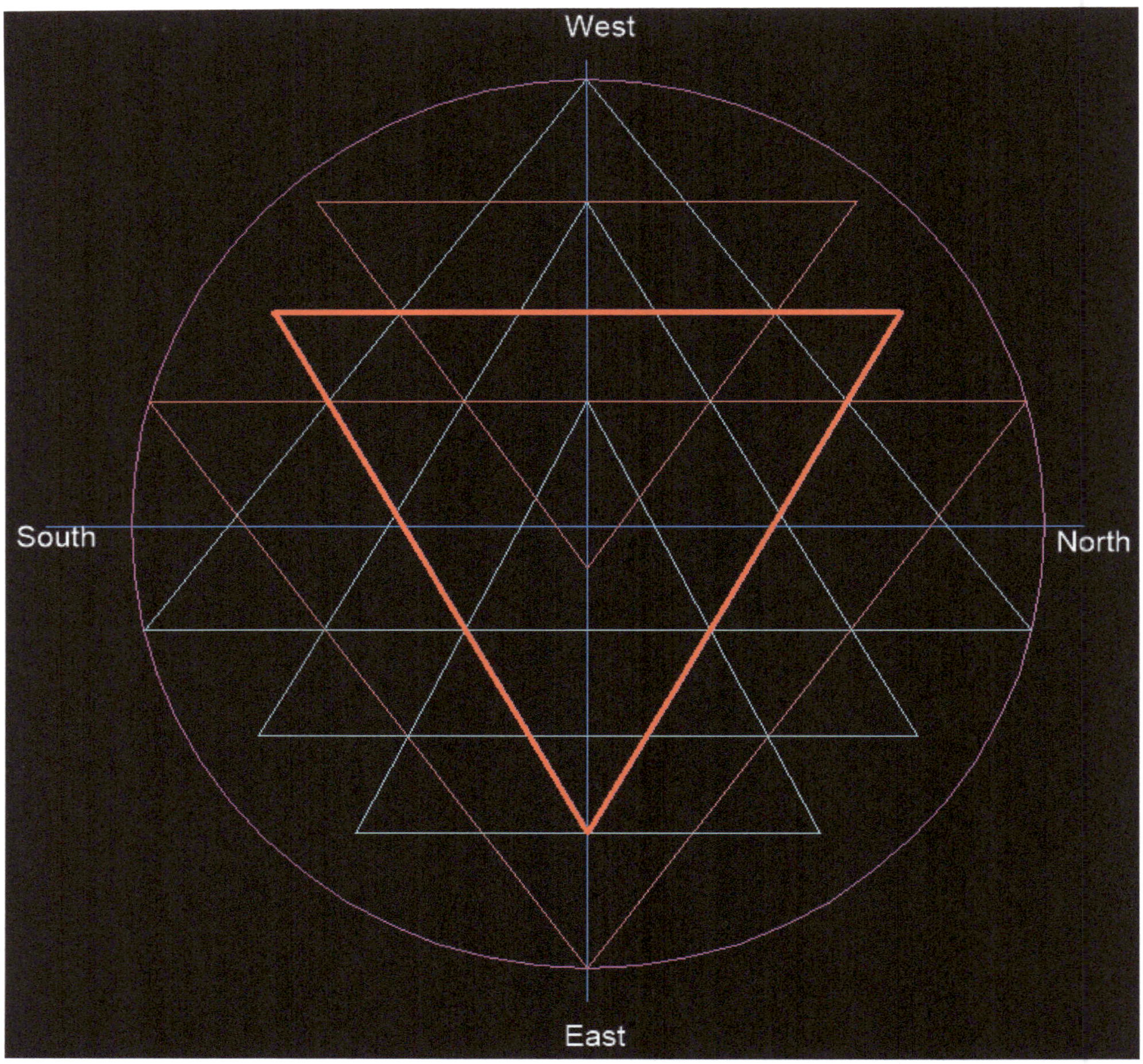

Figure 39 **T2d Completed**

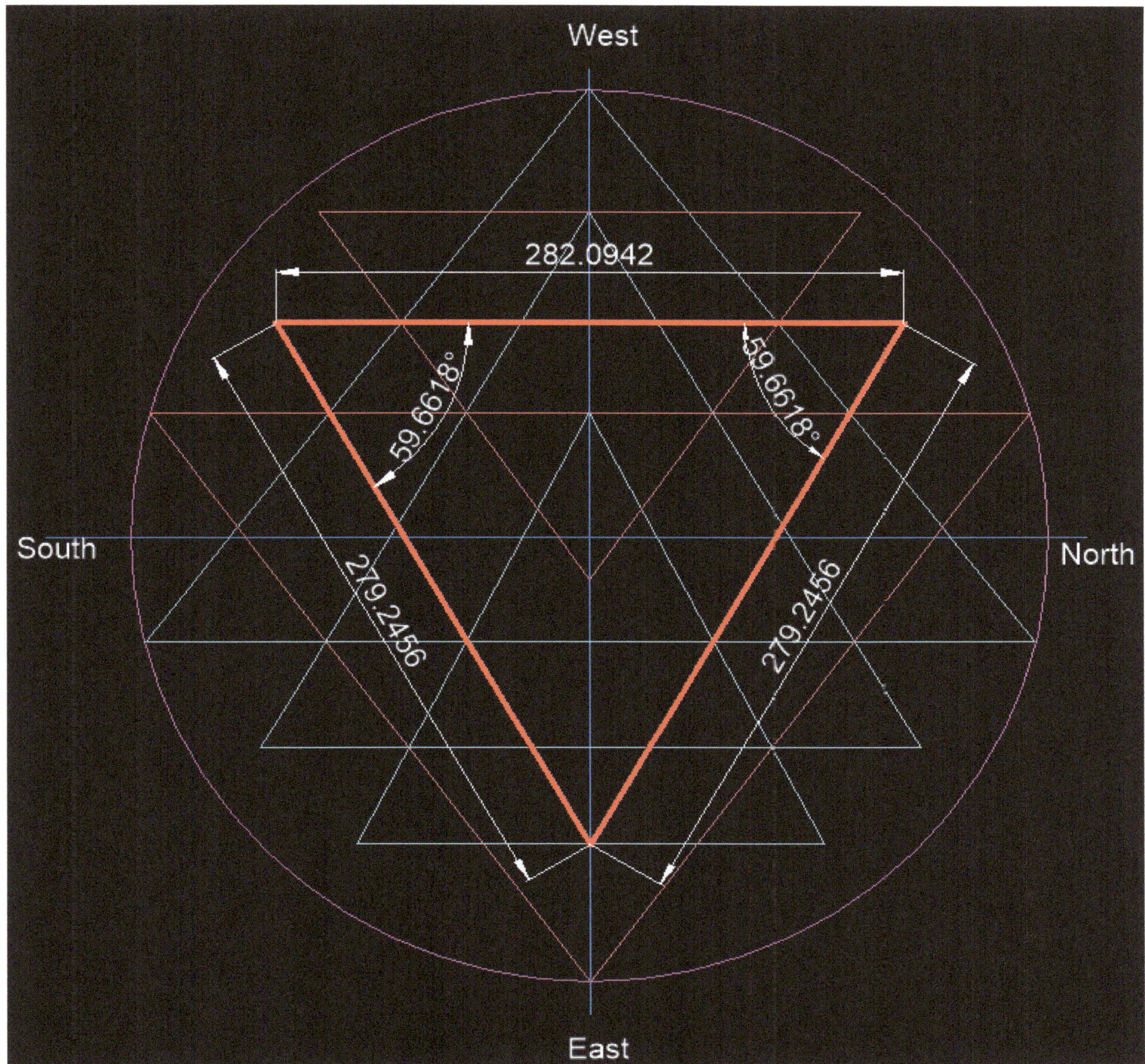

Figure 40 **T2d Dimensions**

T2d dimensions, Isosceles sides = 279.2456 each, Isosceles Angles = 59.6618° each, base = 282.0942 mm. Area = 33993.1765 mm²

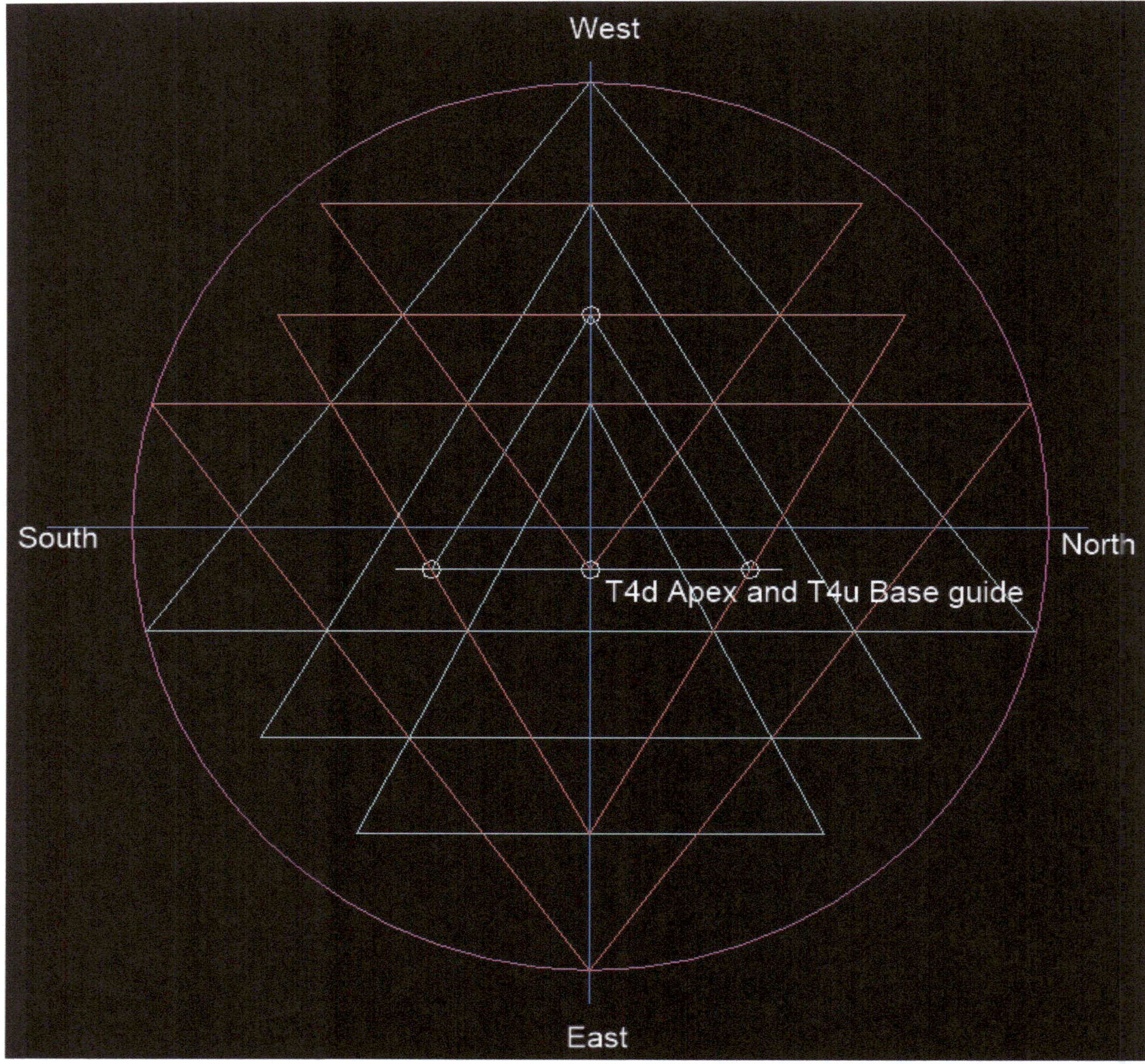

Figure 41 **Finding T3u**

1. We already have the T3u base.
2. The intersection of T2d base with the cross hair vertical gives us the T3u Apex.

38 Complete the fifth Triangle T3u

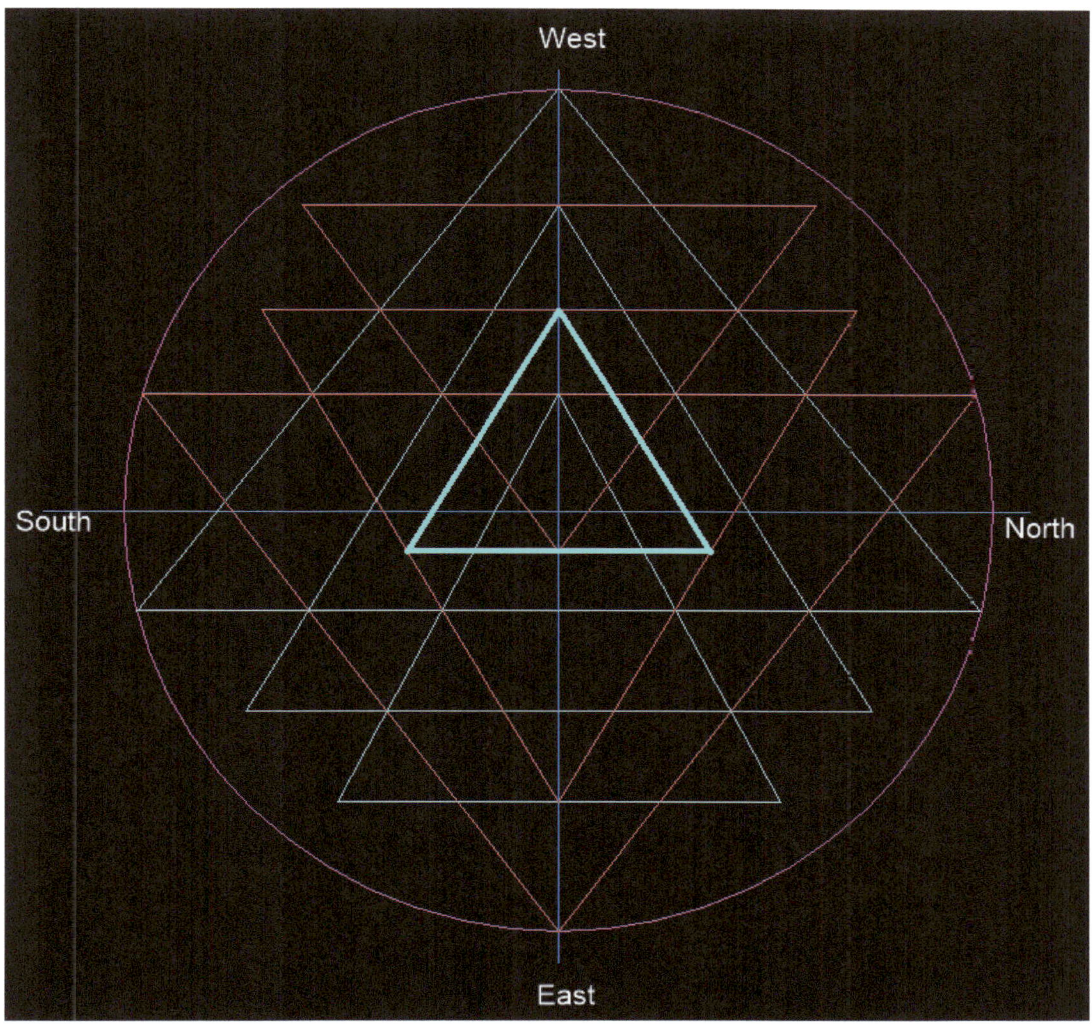

Figure 42 **T3u Completed**

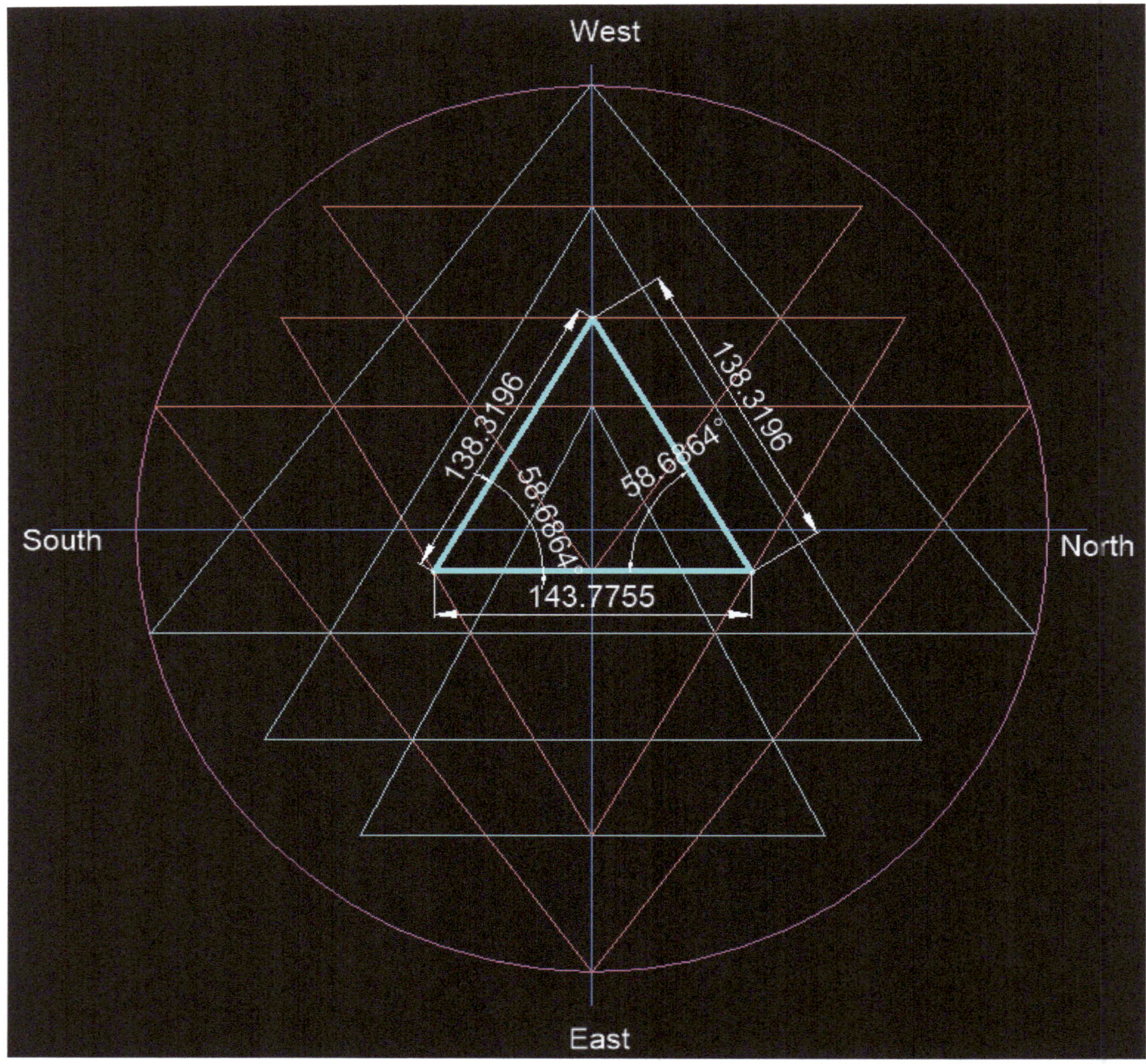

Figure 43 **T3u Dimensions**

T3u dimensions, Isosceles sides = 138.3196 each, Isosceles Angles = 58.6864° each, base = 143.7755 mm. Area = 8495.0710 mm²

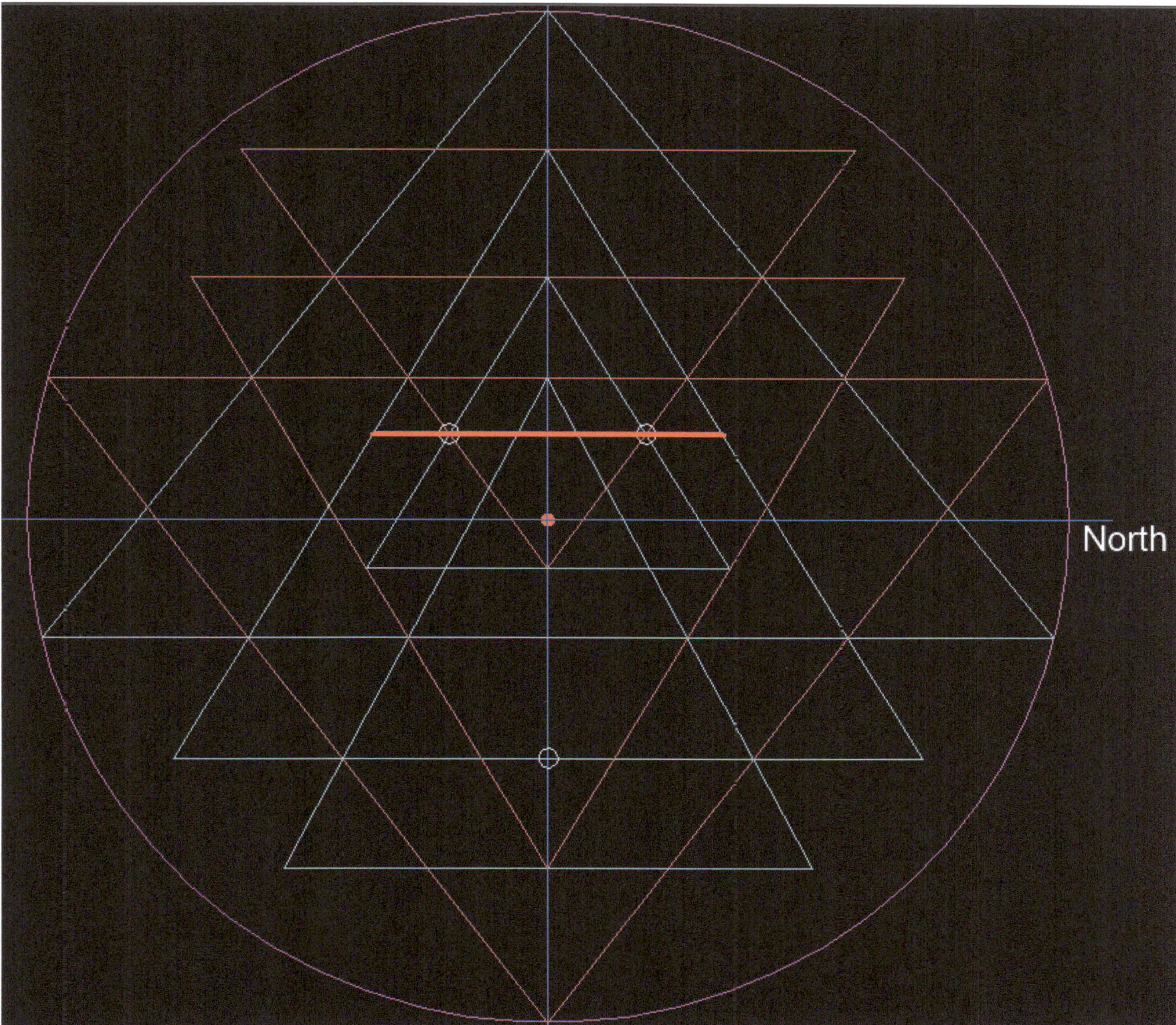

Figure 44 **Finding T3d**

1. The intersections of T3u and T4d give us the guides for T3d base.
2. The Apex for T3d is given by the intersection of T2u base with the Cross Hair vertical.

41 Complete the sixth Triangle T3d

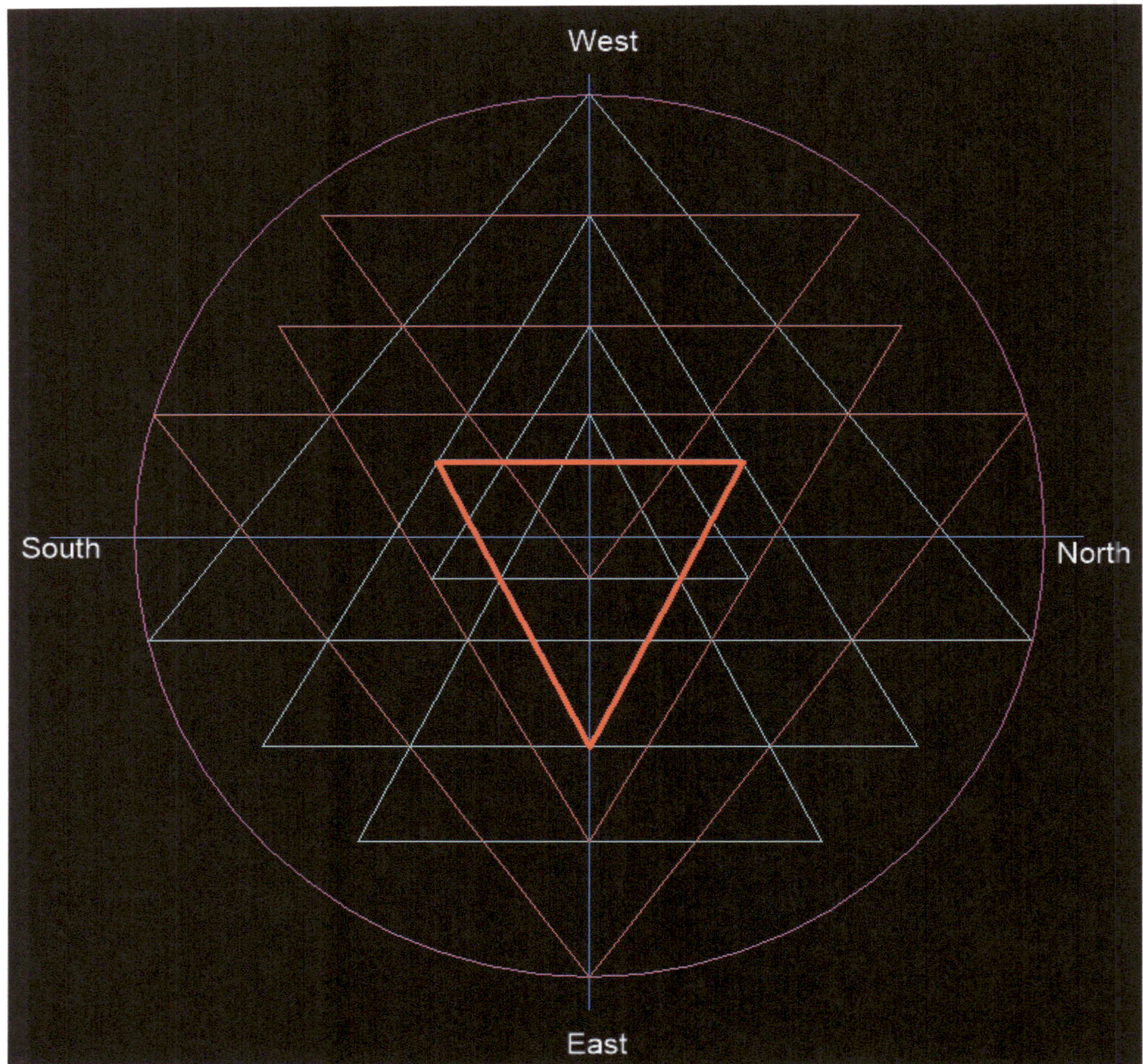

Figure 45 **T3d Completed**

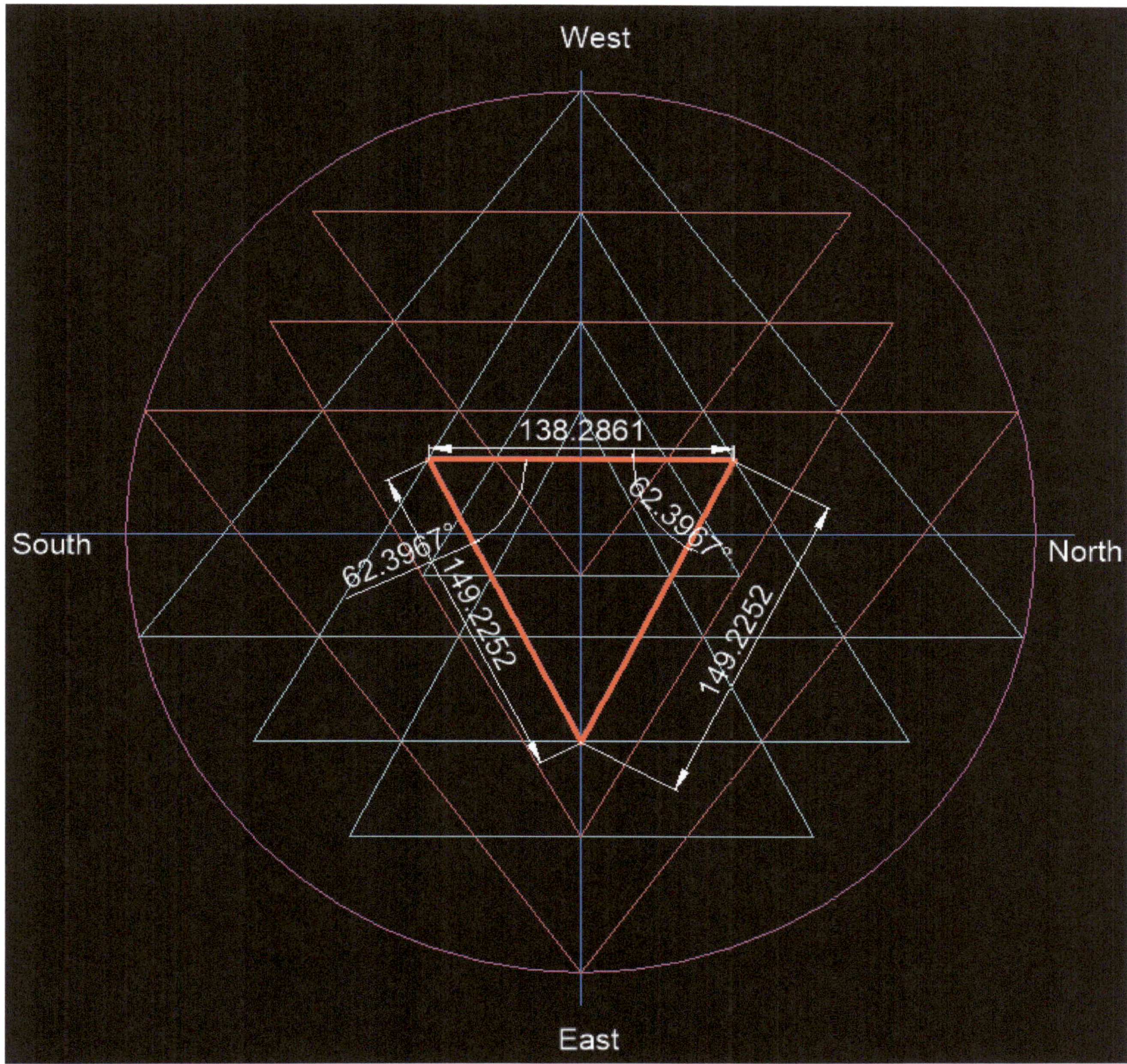

Figure 46 **T3d Dimensions**

T3d dimensions, Isosceles sides = 149.2252 each, Isosceles Angles = 62.3967° each, base = 138.2861 mm. Area = 9143.4720 mm²

43 Guides for sides of ninth Triangle T5d

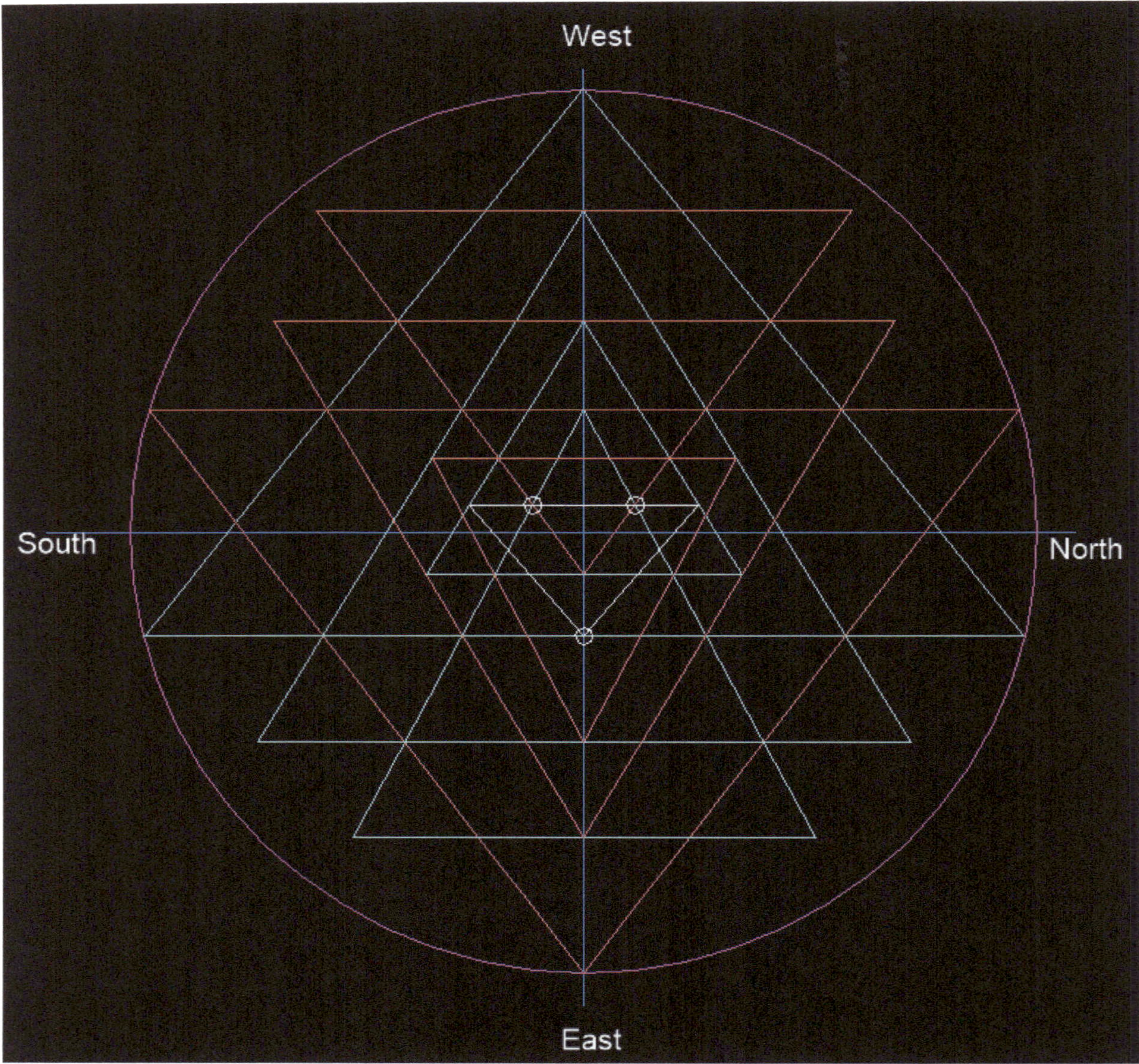

Figure 47 **Finding T5d**

1. The intersections of T4u and T4d give us the guides for T5d base.
2. The Apex for T5d is given by the intersection of T1u base with the Cross Hair vertical.

44 Complete the ninth Triangle T5d

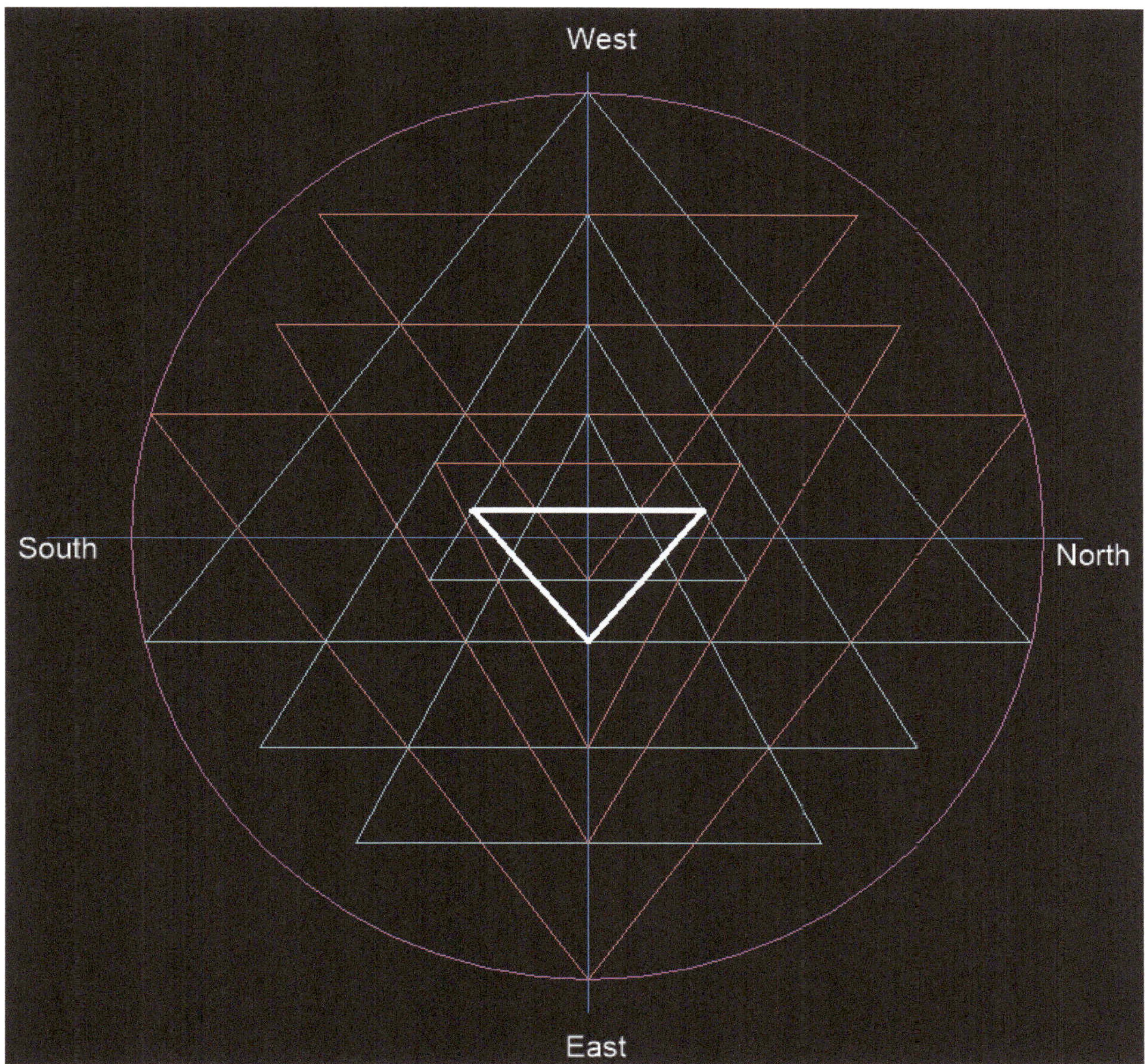

Figure 48 **T5d Completed**

45 Dimensions of ninth Triangle T5d

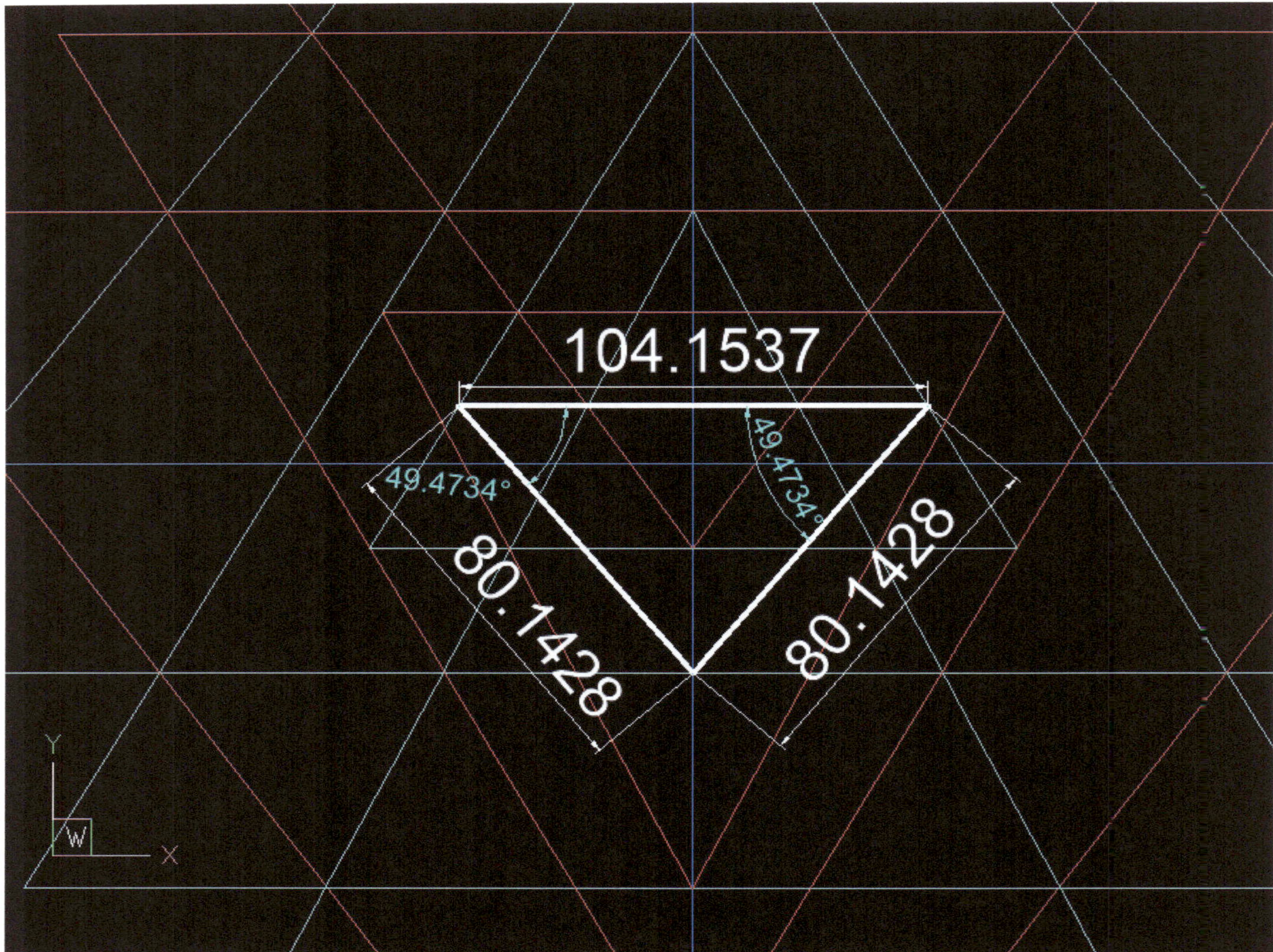

Figure 49 **T5d Dimensions**

T5d Baseline = 104.1537mm. Half baseline = 104.1537/2 = 52.07685mm, Sides = (Hypotenuse) = 80.1428mm.
Hypotenuse/Half baseline = 80.1428/52.07685 = **1.5389**, Isosceles Angles = 49.4734° each.

It is close but not quite an equilateral triangle. It happens to be an isosceles triangle.

46 Finding the center of T5d for Bindu

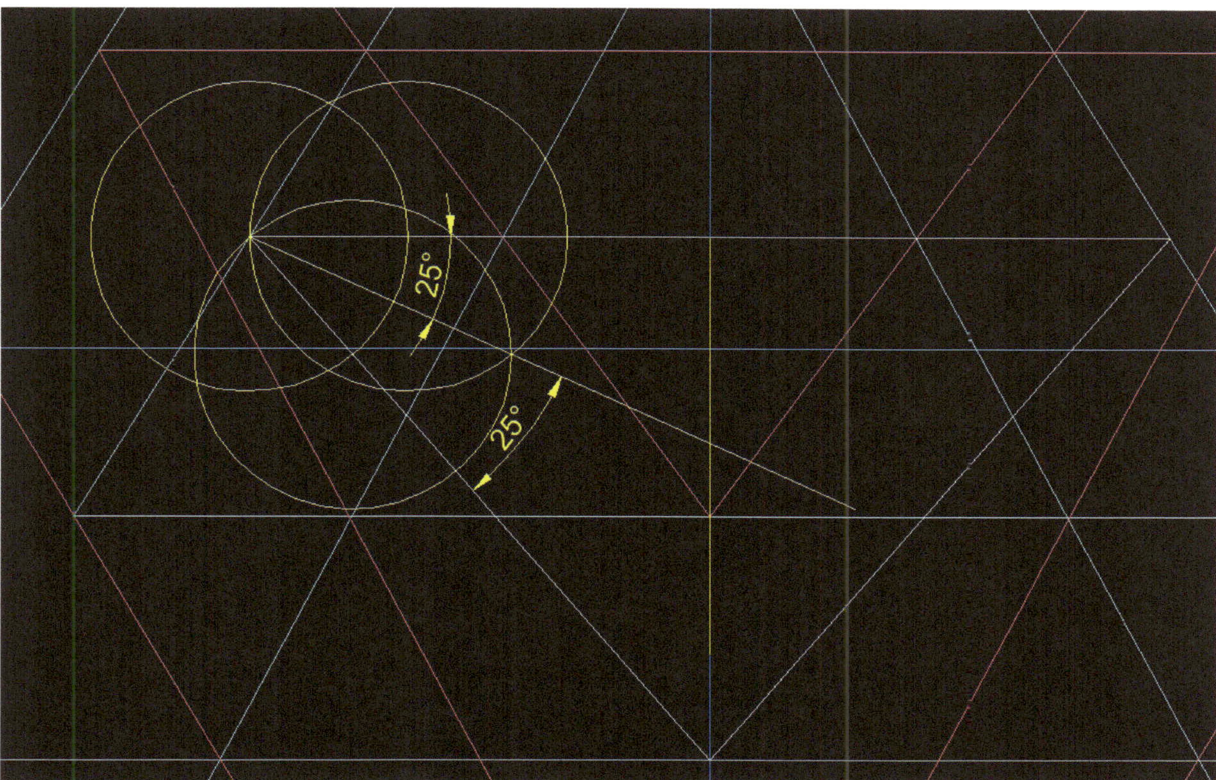

Figure 50 **Finding Bindu**

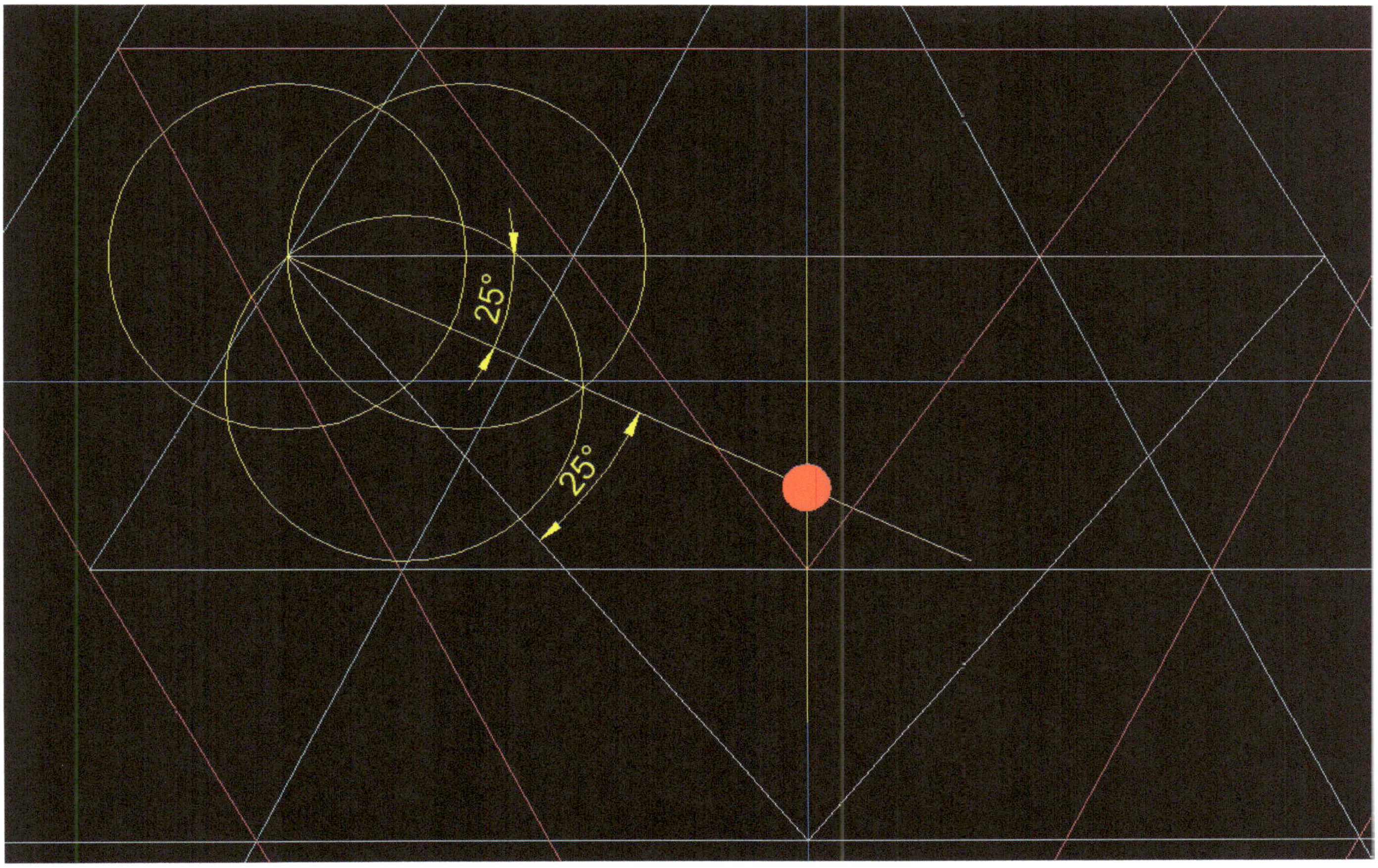

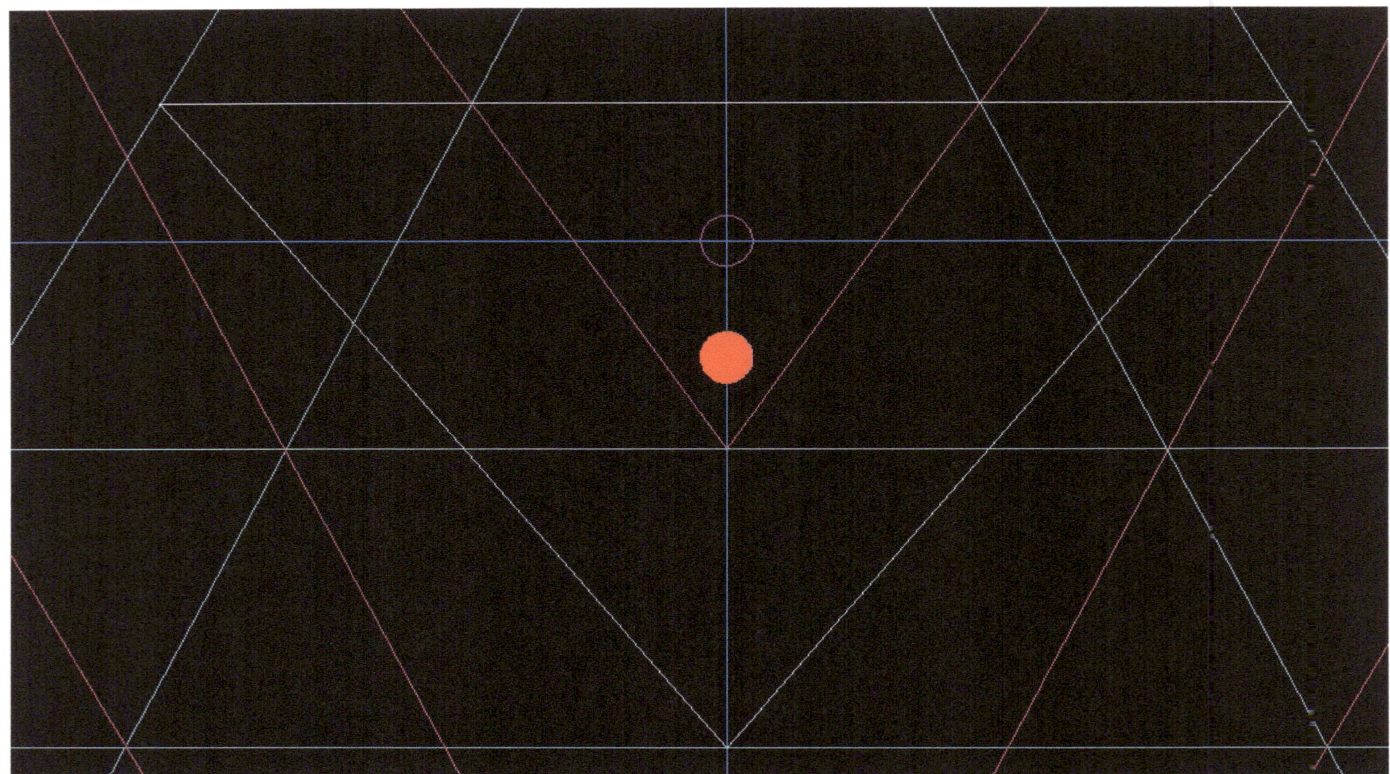

Using the angle bisectors for T5d, we obtain the intersection point, that gives us the center for drawing a new inscribed circle. This new circle is filled in with red color by hatching and known as the Bindu.

Size of the Bindu can be kept relatively small, as it represents the infinitesimal Union of the Divine with the Creation.

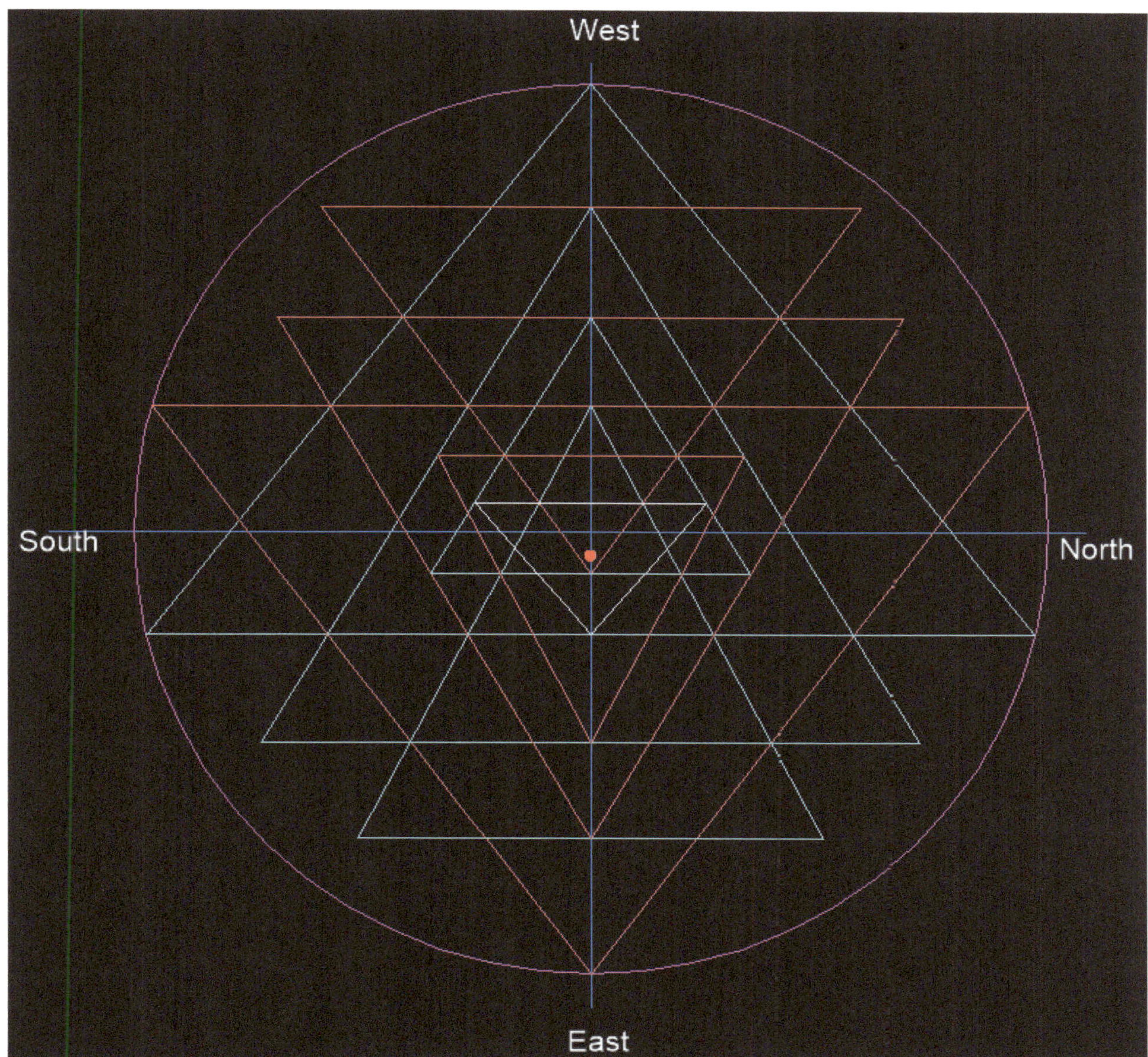

Figure 51 **Bindu Completed**

47 Sri Yantra core of 9 Basic Triangles

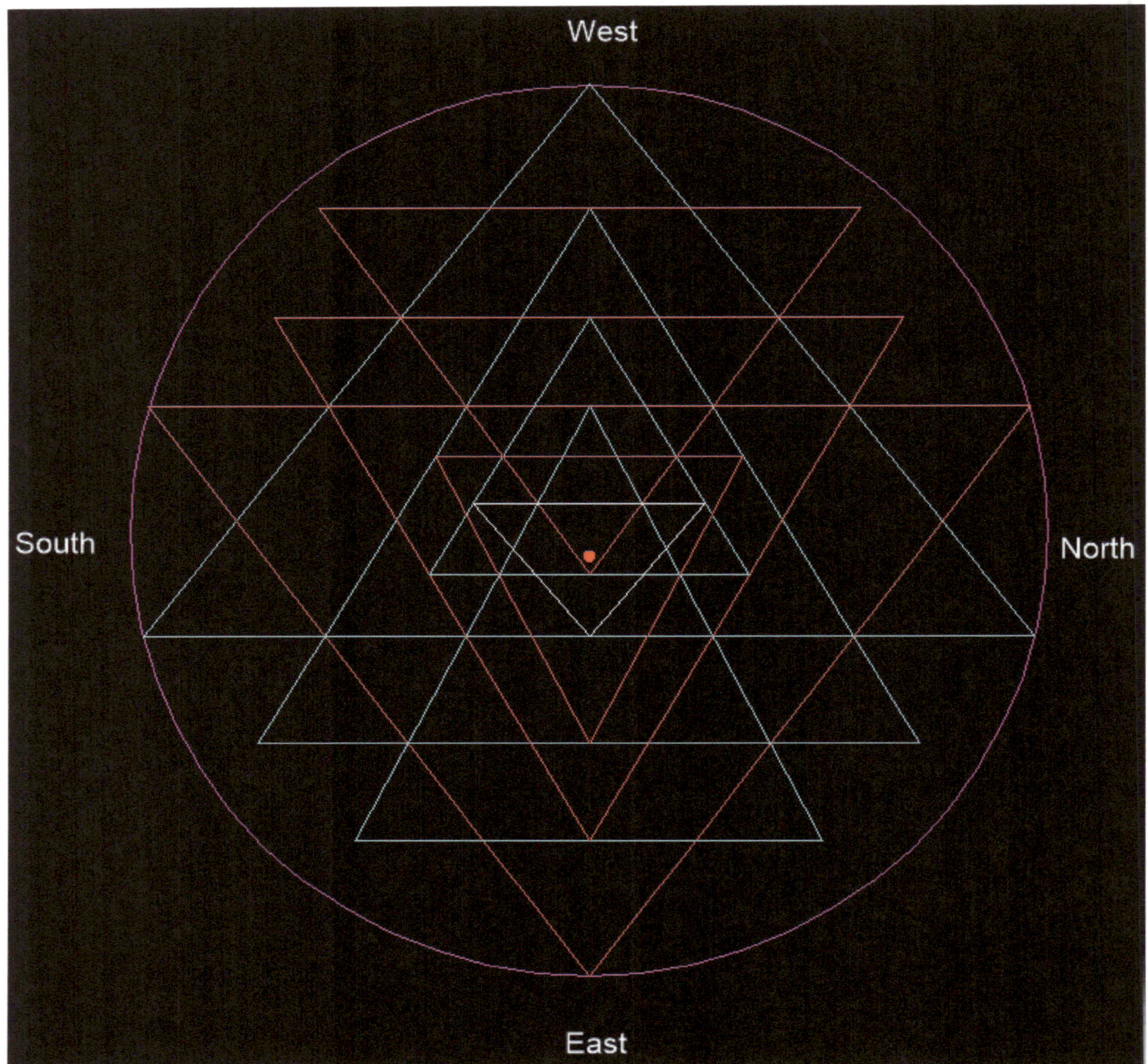

Sri Yantra core consists of
- a **Circle** encircling the two largest triangles
- **9 basic Triangles**
- a **Bindu** red dot in the center.
- the **Direction** vector is also important, with **East** being below.

T1u Golden Ratio Triangle.
T1u dimensions, isosceles side = 323.6068 each, base = 400. Isosceles Angles = 51.82729° each.
Ratio of isosceles side/half base = 323.6068/200 = **1.618034** = Φ Phi

Dimensions of Square Rectangle 3Circles Golden Ratio Triangle

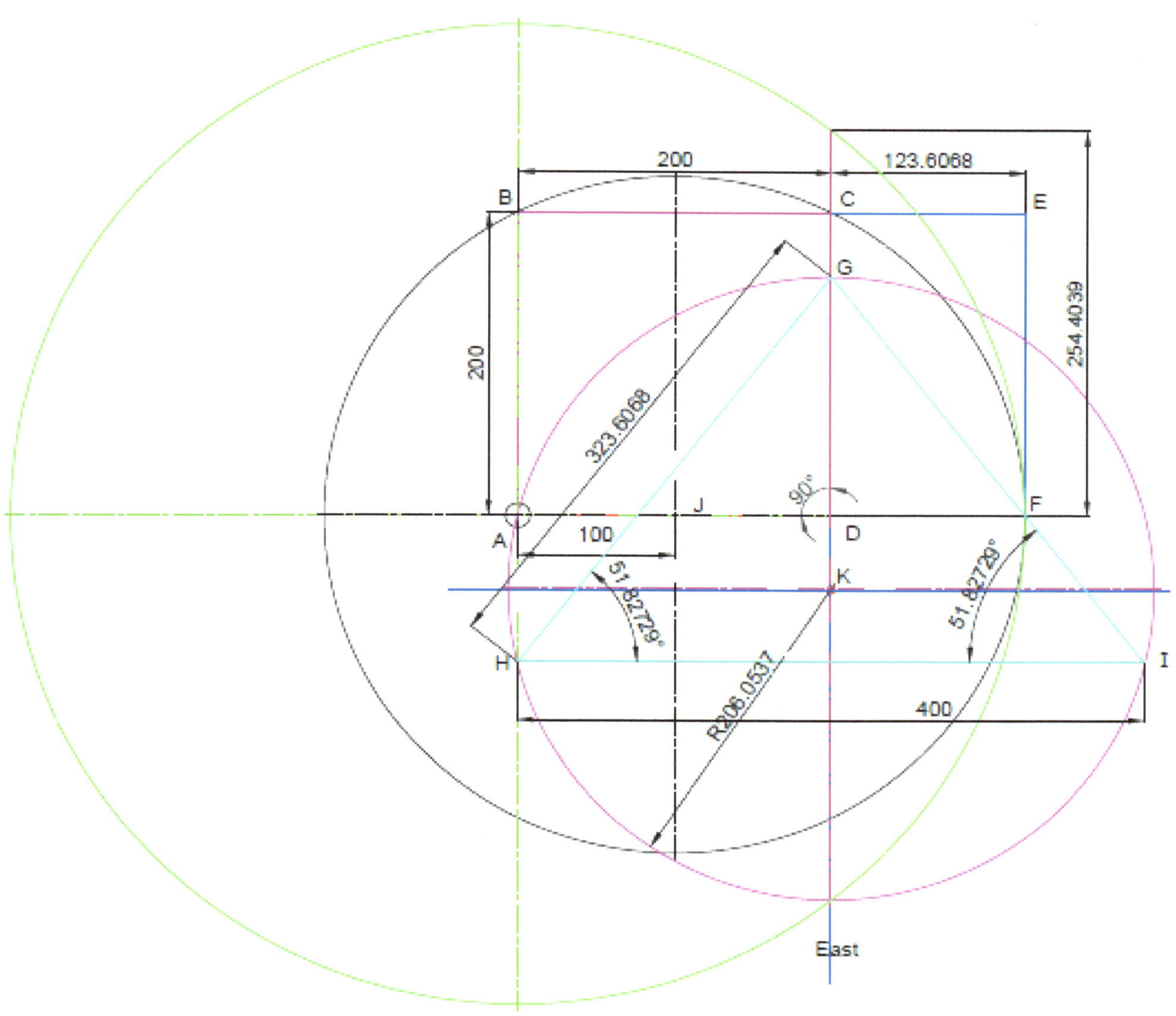

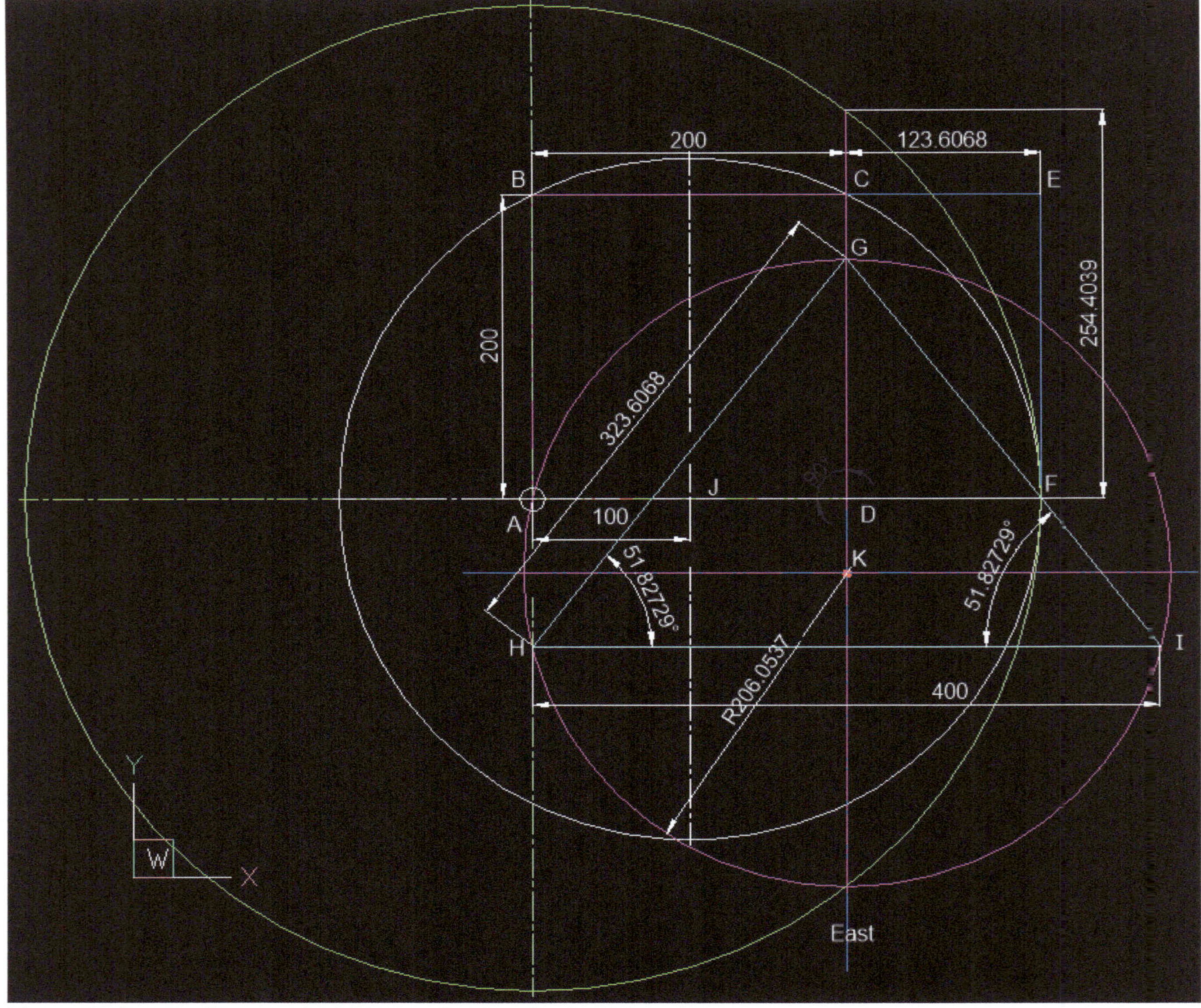

Figure 53 **Square Rectangle 3Circles Golden Ratio Triangle**

1. Square = **ABCD** of side 200mm
2. Right Angle Triangle = JDC, base JD = 100, hypotenuse JC = 223.6068
3a. Golden Ratio Rectangle = **ABEF**, shorter side AB = EF = 200, longer side BE = AF = **323.6068**
3b. Golden Ratio Rectangle = **ABEF**, ratio of longer side/shorter side = 323.6068/200 = **1.618034** = ϕ Phi
4. Black Circle with center J has radius = hypotenuse JC = 223.6068
5. Green Circle with center A has radius = longer side of Rectangle AF = 323.6068
7a. **Apex** of Golden Ratio Triangle = G
7b. Golden Ratio Triangle **T1u** = Largest Upward Apex Triangle = GHI
7c. T1u Dimensions, base HI = 400, sides GH = GI = **323.6068**, Isosceles Angles GHI = GIH = **51.8273°**
8. Pink Circle with center K has radius = **206.0537**

Sri Yantra Accuracy and Precision

A Sri Yantra designed with accuracy and precision is precious, beautiful and highly benevolent.

In its basic form, a Sri Yantra consists of 4 Upward Apex Triangles, 5 Downward Apex Triangles, an encircling circle and the Bindu. Some points to be noted:

- **ONLY** the largest Upward Apex Triangle has Golden Ratio proportions of
    - Two angles of 51.8272° each.
    - Triangle half-baseline length = 1 unit. (i.e., baseline length ratio = 2 units)
    - Triangle two sides length each = Φ Phi = 1.618 units.
    - Triangle bisect vertical length = $\sqrt{\Phi}$ = 1.272 units.
- The largest Downward Apex Triangle is an isosceles triangle with angles of 53°, near to but not exactly Golden Ratio.
- **Bindu** lies on the vertical crosshair, and just below the horizontal crosshair. It is not in the exact center of the circle.
- **Smallest inner** downward apex triangle encircling Bindu must be close to an equilateral triangle.
- There are 18 **Marma** points, i.e., 3 lines intersecting junction.
- There are 24 **Sandhi** points, i.e., 2 lines intersecting junction.
- A maximum of **two** Marma junctions that have an **error** of all 3 lines not meeting precisely but creating a small triangle.
- All **3 tips** of the largest Upward Apex Triangle and the largest Downward Apex Triangle should **touch** the circle, without getting bisected or chopped off by the circle. No other triangle should touch the circle.

For ease in understanding, we give arbitrary names to the triangles.
T1u = Upward Apex Largest Triangle, having Golden Ratio.
T2u, T3u, T4u are the other Upward Apex Triangles in sequence from top.
T1d, T2d, T3d, T4d are the four Downward Apex Triangles in sequence from below.
T5d downward apex triangle is the innermost triangle enclosing the Bindu.

Upward Apex Golden Ratio Triangle T1u

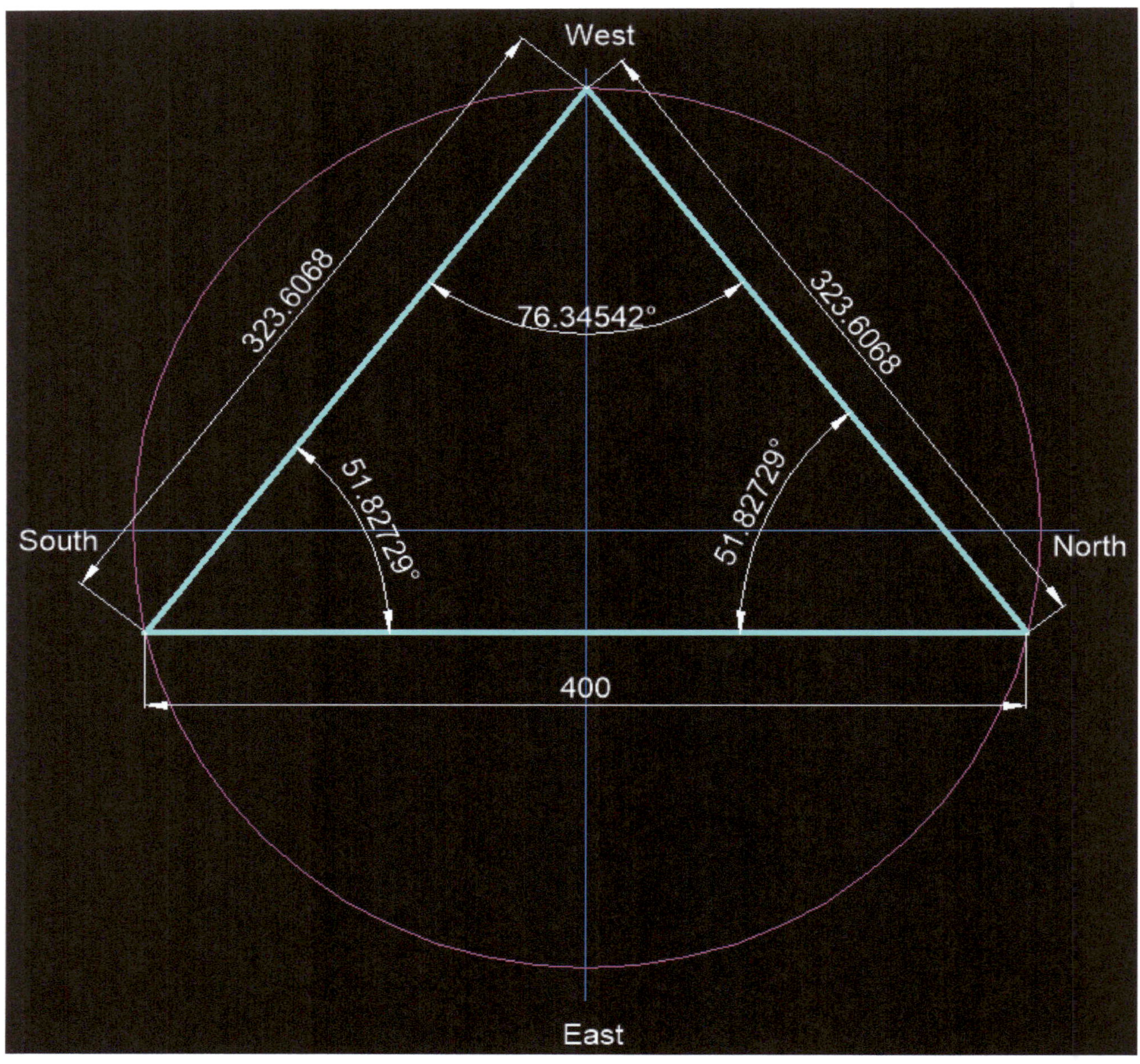

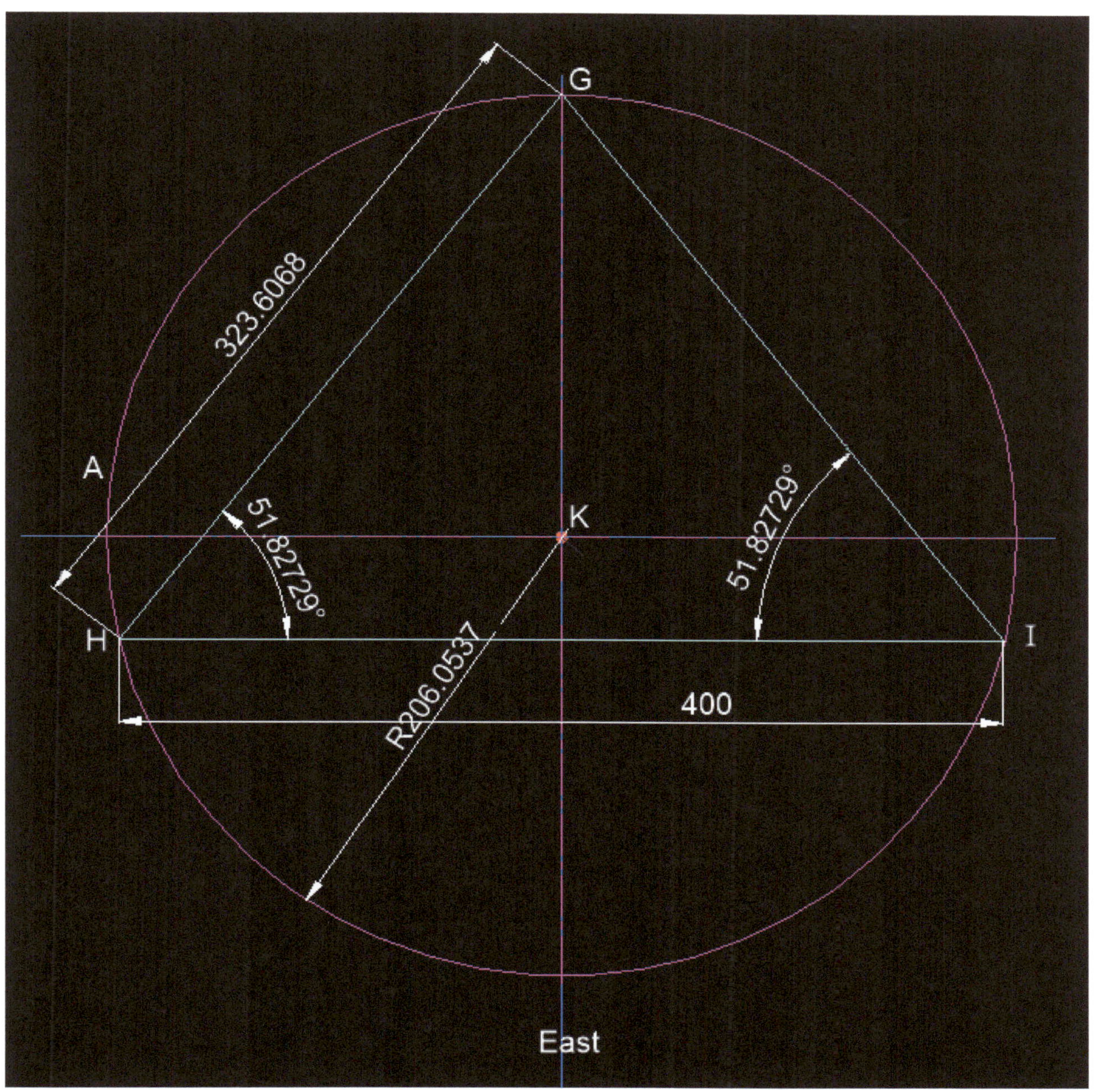

G
A
K
H
I
323.6068
51.82729°
51.82729°
R206.0537
400
East

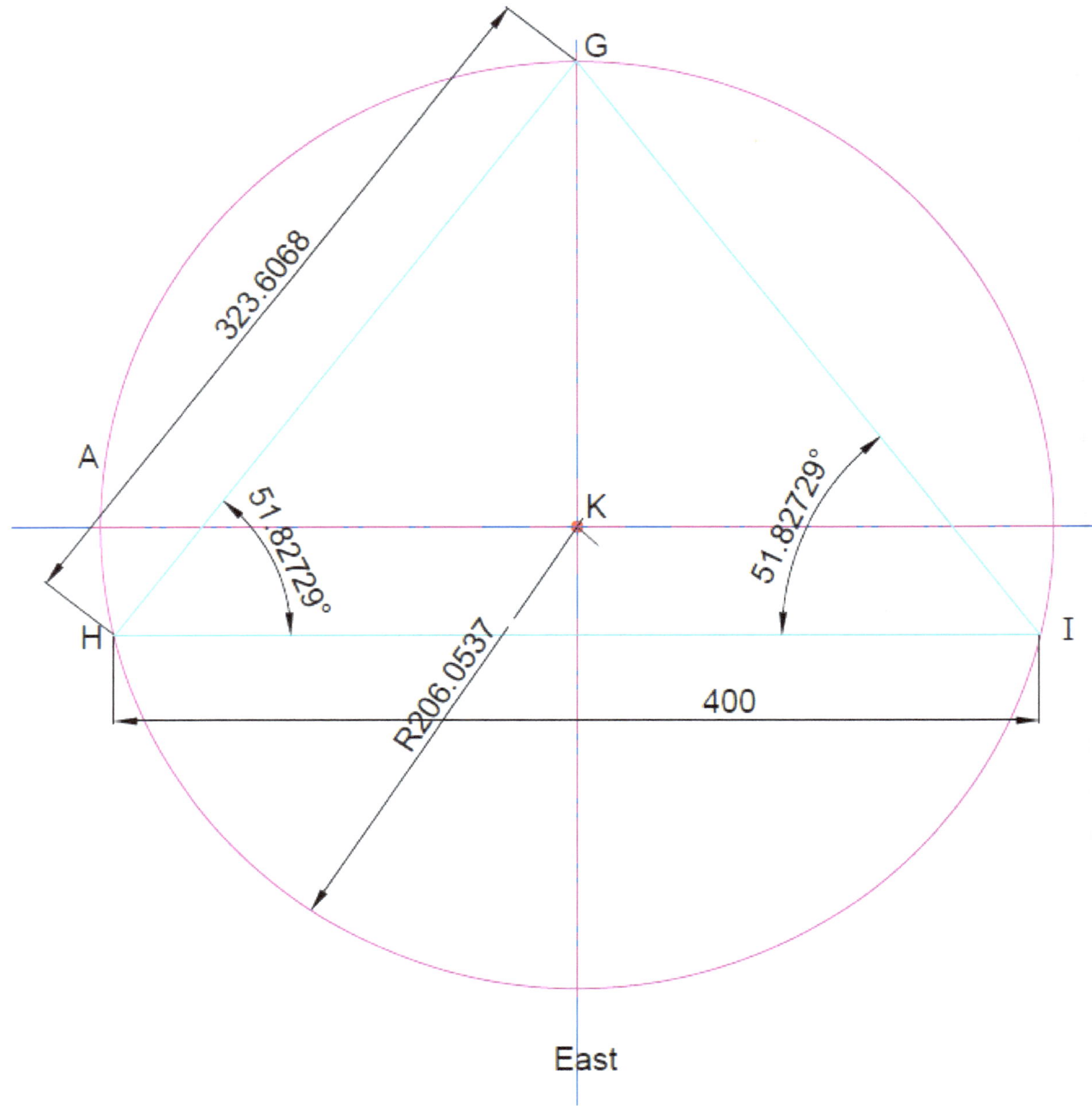

Figure 54 **T1u Dimensions are as per Golden Ratio**

Baseline = 399.9981mm. Half baseline = 399.9981/2 = 199.999mm
Sides = (Hypotenuse) = 323.6068mm each.
Hypotenuse/Half baseline = 323.6068/199.999 = **1.6180** = Φ Phi Golden Ratio.
Vertical length = 254.4mm and Vertical/Half baseline = 254.4/199.999 = **1.272** = 1/1.6180 = 1/Φ
Isosceles Angle = 51.8273°

Downward Apex Largest Triangle T1d

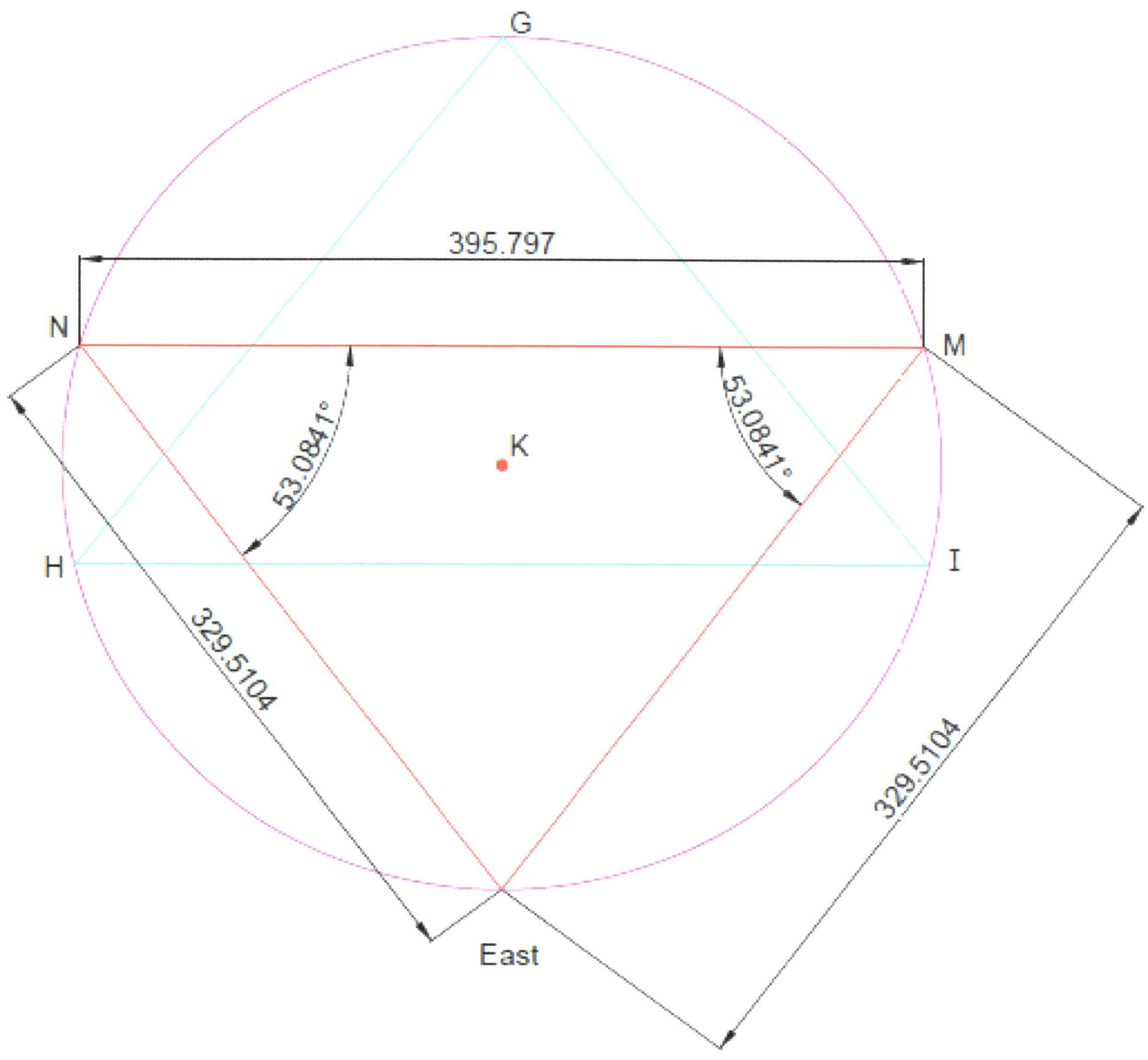

Figure 55 **T1d Dimensions not in Golden Ratio**

Baseline = 395.1744mm. Half baseline = 395.1744/2 = 197.5872mm
Sides = (Hypotenuse) = 329.312mm each.
Hypotenuse/Half baseline = 329.312/197.5872 = **1.6667**.
Isosceles Angle = 53.1301°

Core Complete

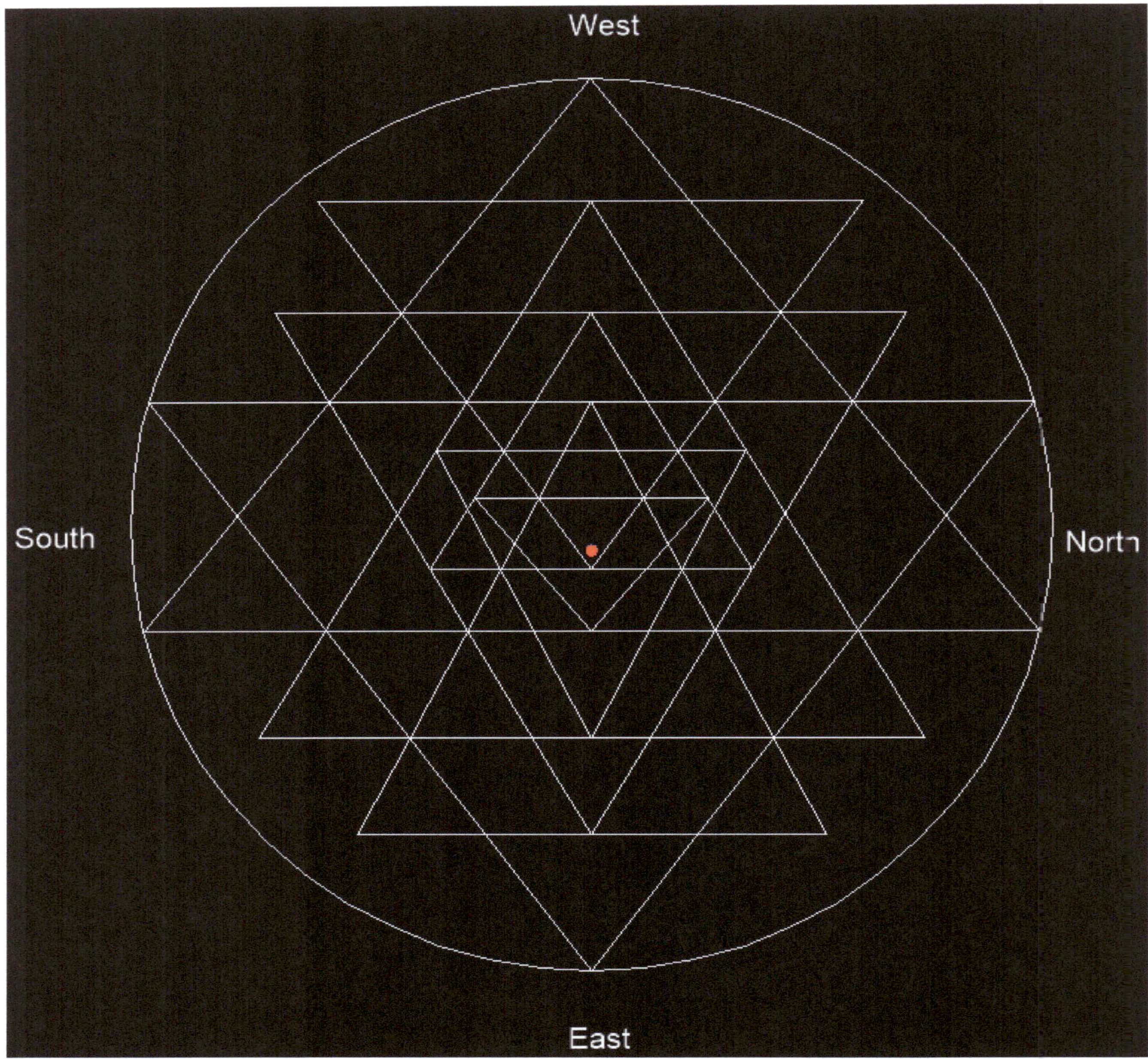

Figure 56 **Triangles**

Downward Apex Innermost Triangle T5d

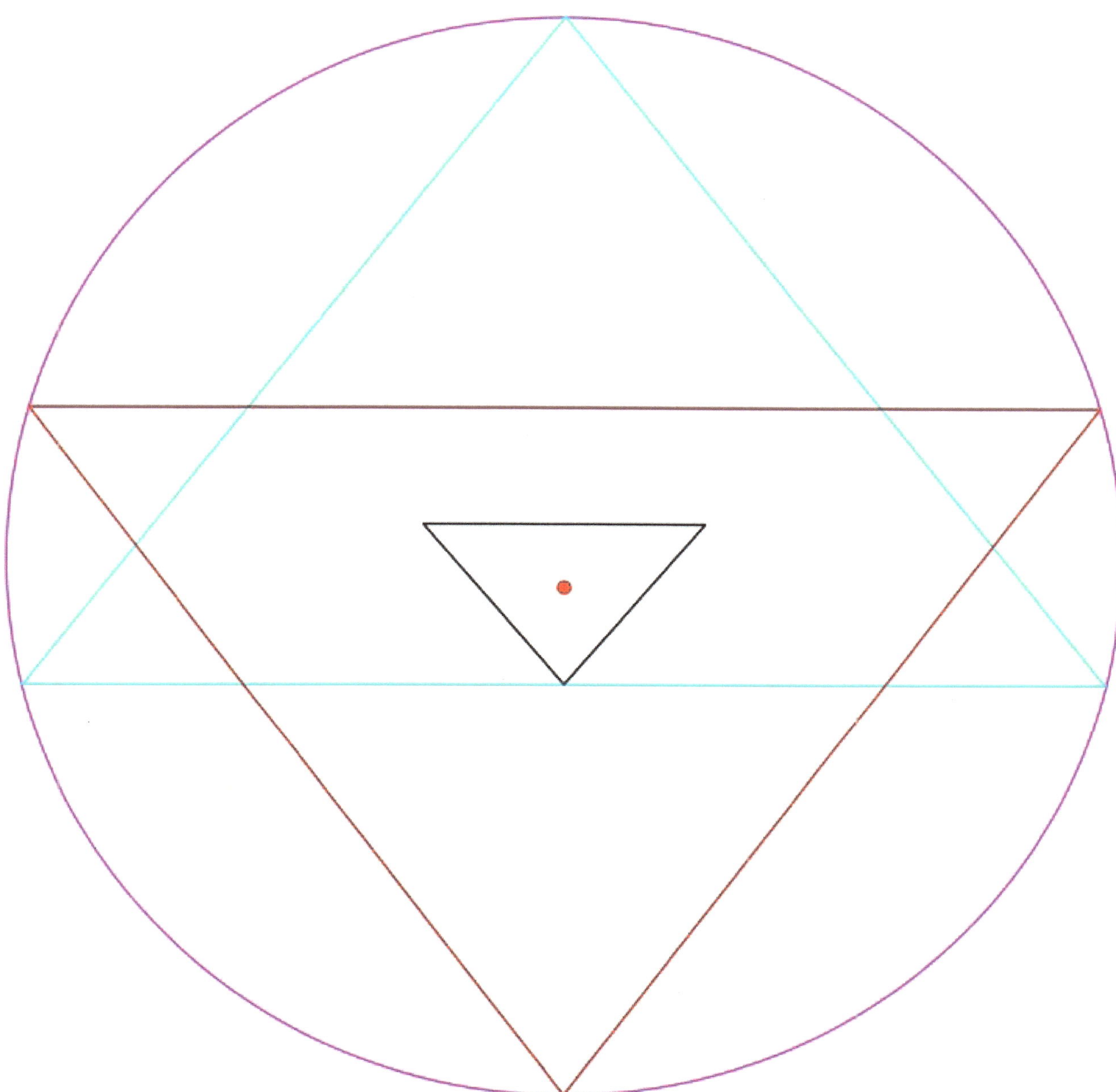

Figure 57 **T5d Triangle Innermost**

i Triangle T1u Golden Ratio

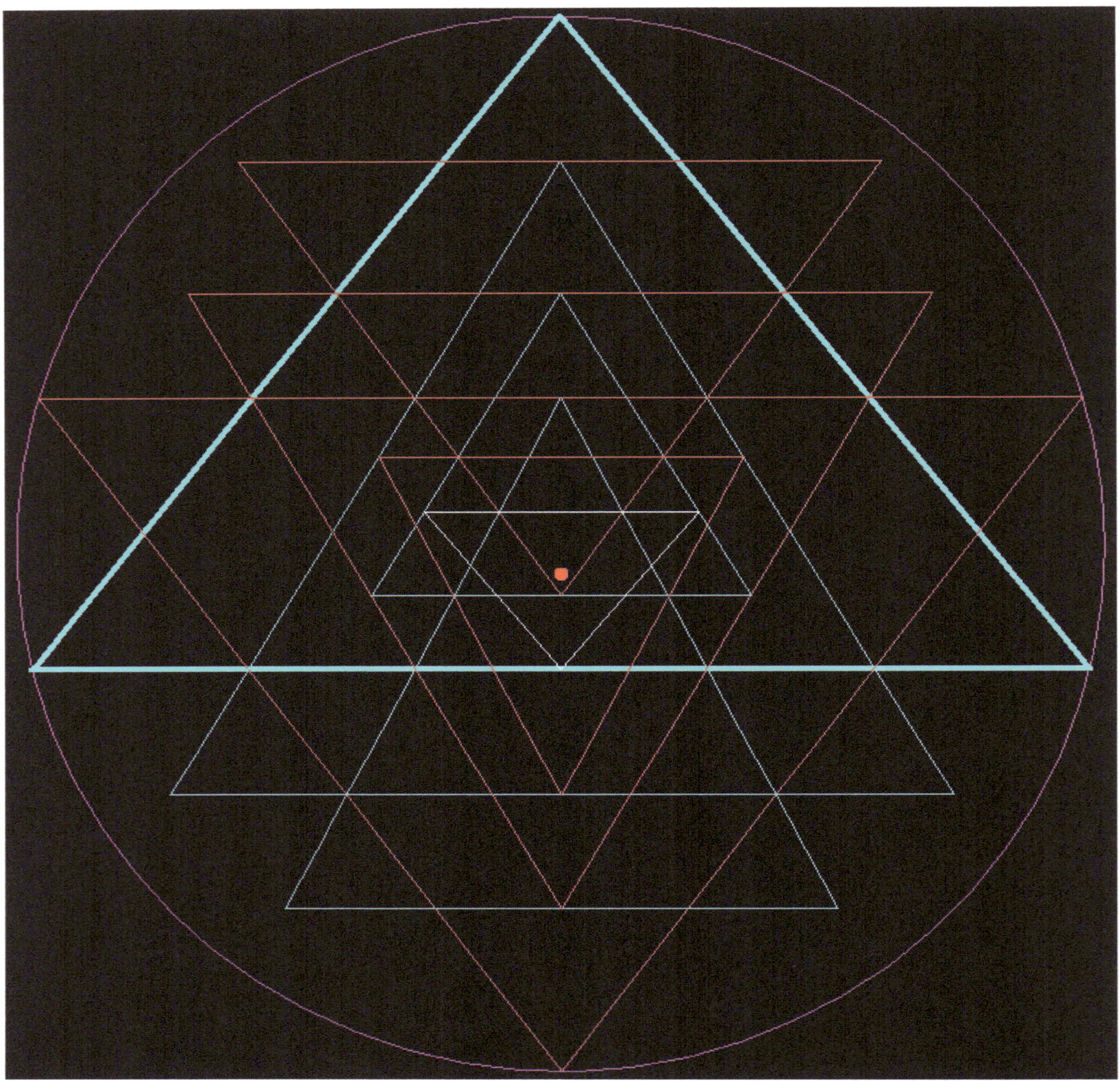

T1u Dimensions, Isosceles sides = 323.6068 each, Base = 400, Isosceles Angles = 51.82729° each. Area 50880.78655 mm²

ii Triangle T1d

T1d dimensions, Isosceles sides = 329.5104 each, Isosceles Angles = 53.0841° each, base = 395.797 mm. Area = 52139.1334 mm²

iii Triangle T2u

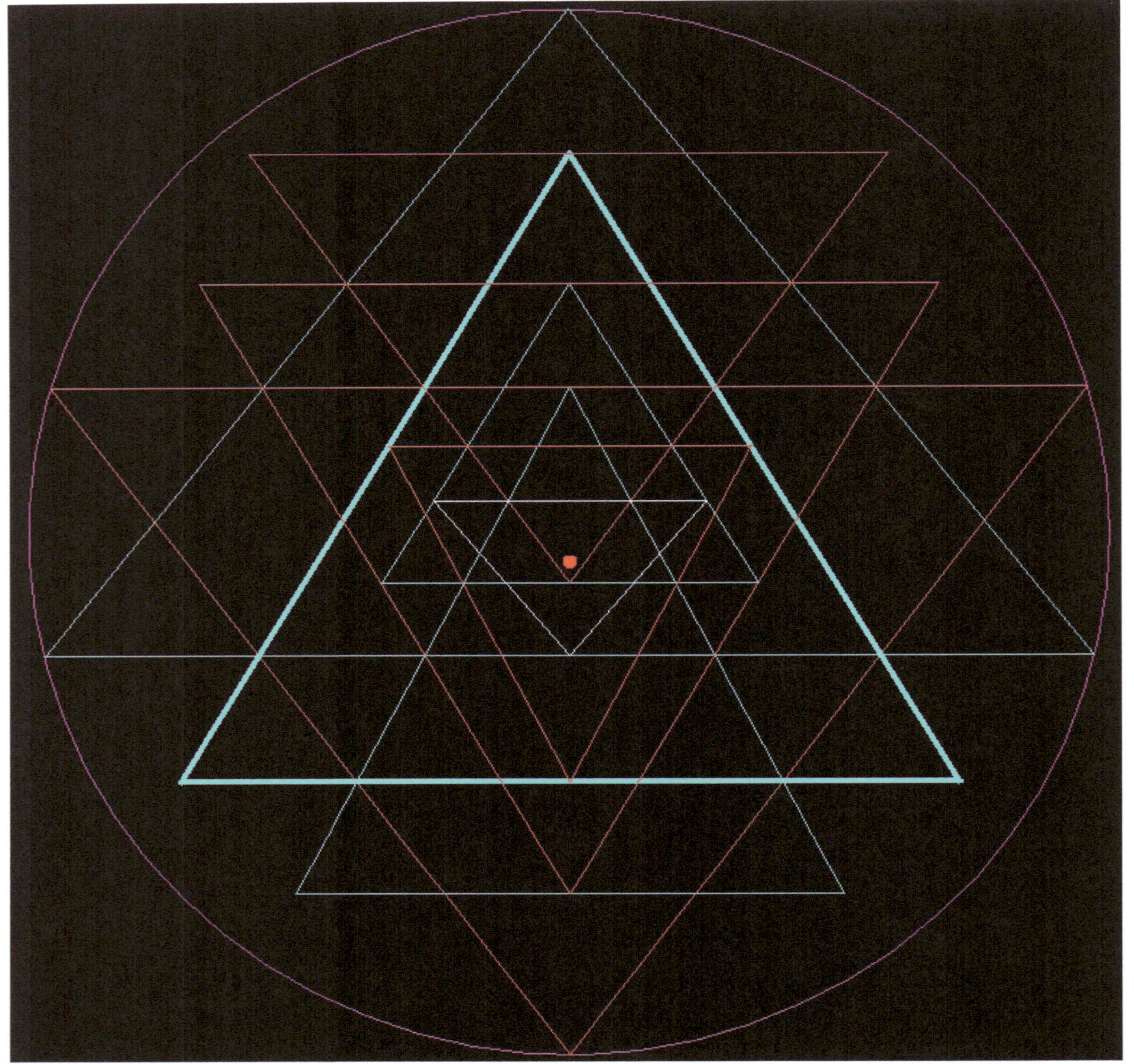

T2u dimensions, Isosceles sides = 288.6457 each, Isosceles Angles = 59.0531° each, base = 296.868 nm. Area = 36745.64175 mm²

iv Triangle T2d

T2d dimensions, Isosceles sides = 279.2456 each, Isosceles Angles = 59.6618° each, base = 282.0942 mm. Area = 33993.1765 mm²

v Triangle T3u

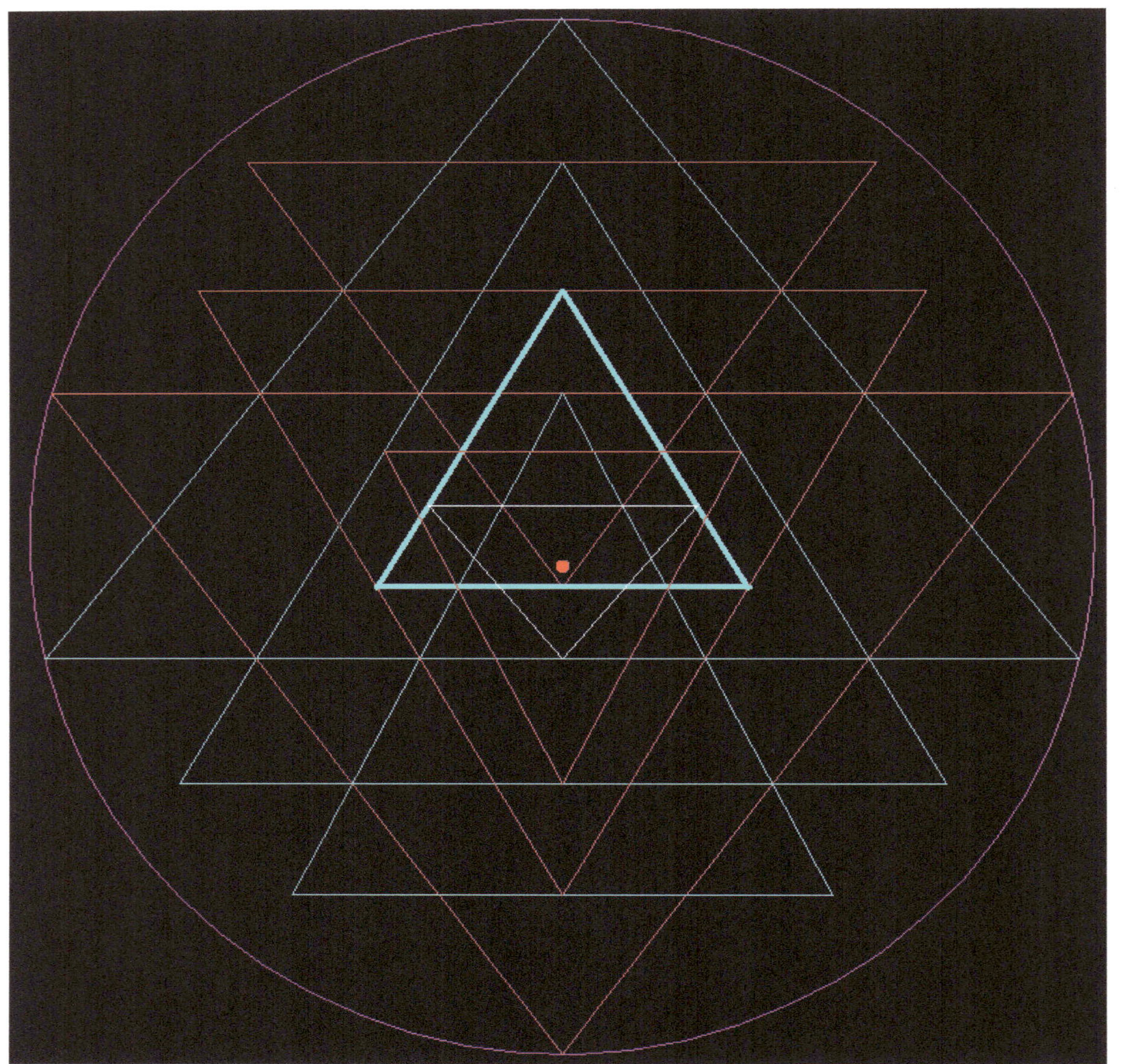

T3u dimensions, Isosceles sides = 138.3196 each, Isosceles Angles = 58.6864° each, base = 143.7755 mm. Area = 8495.0710 mm^2

vi Triangle T3d

T3d dimensions, Isosceles sides = 149.2252 each, Isosceles Angles = 62.3967° each, base = 138.2861 mm. Area = 9143.4720 mm²

vii Triangle T4u

T4u dimensions, Isosceles sides = 329.5104 each, Isosceles Angles = 53.0841° each, base = 395.797 mm. Area = 52139.13345 mm^2

viii Triangle T4d

T4d dimensions, Isosceles sides = 208.6704 each, Isosceles Angles = 54.3331° each, base = 243.3395 mm. Area = 20626.45269 mm²

ix Triangle T5d contains Bindu

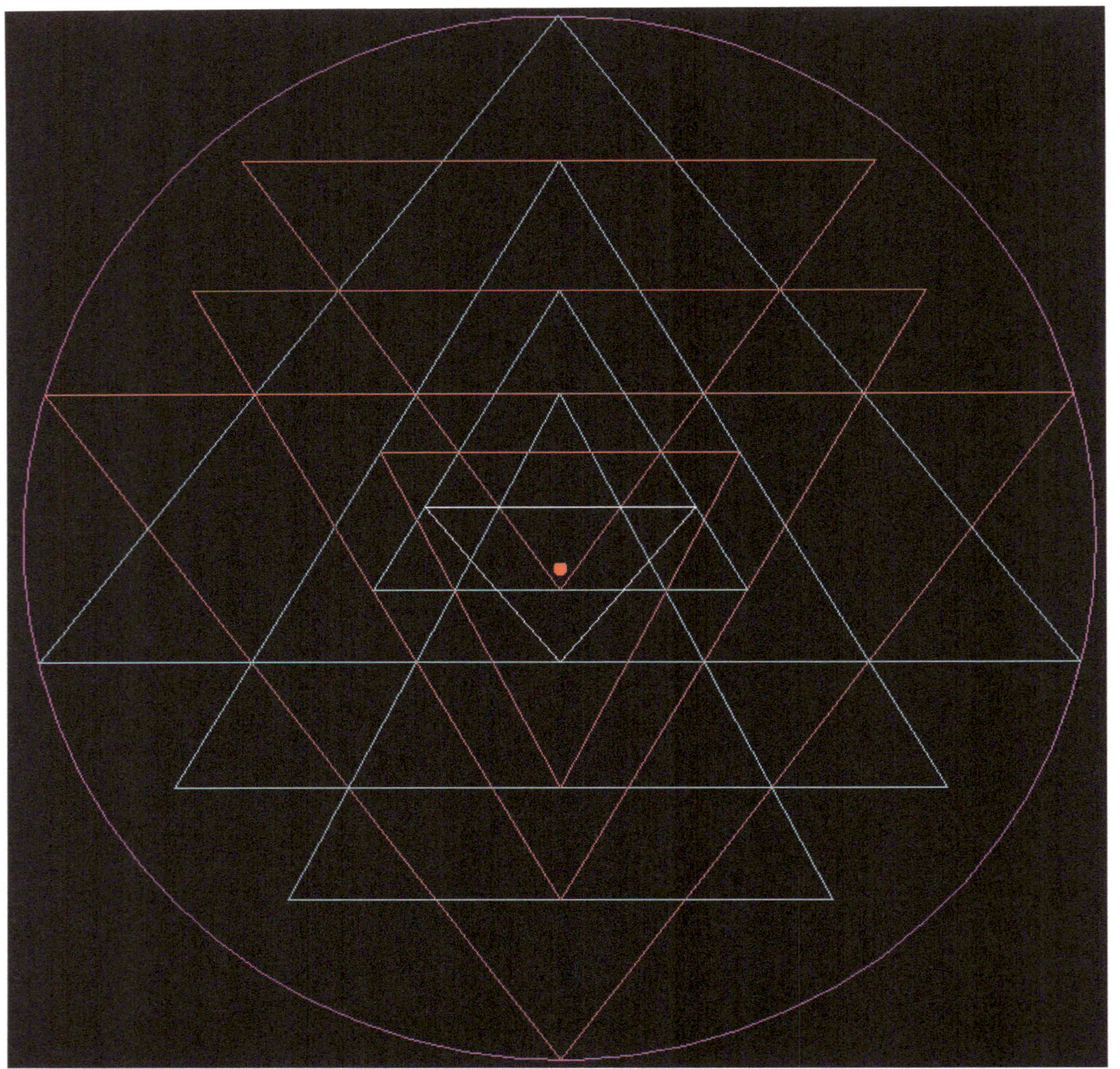

Figure 66 **T5d contains Bindu**

T5d dimensions, Isosceles sides = 80.1428 each, Isosceles Angles = 49.4734° each, base = 104.1537 mm. Area = 3172.36085 mm²

Four Upward Apex Shiva Triangles

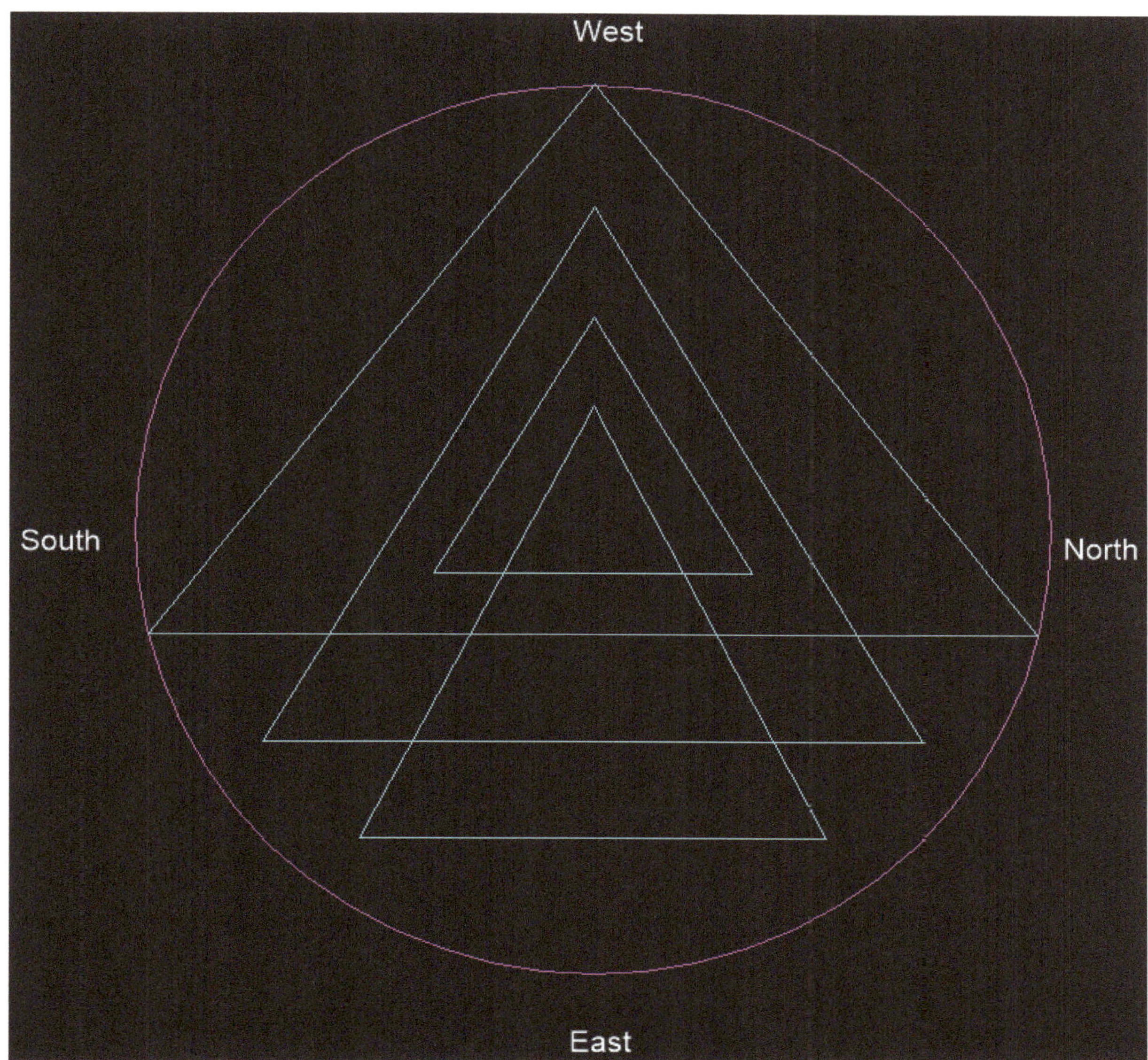

Figure 67 **Four Upward Apex**

Four Downward Apex Shakti Triangles

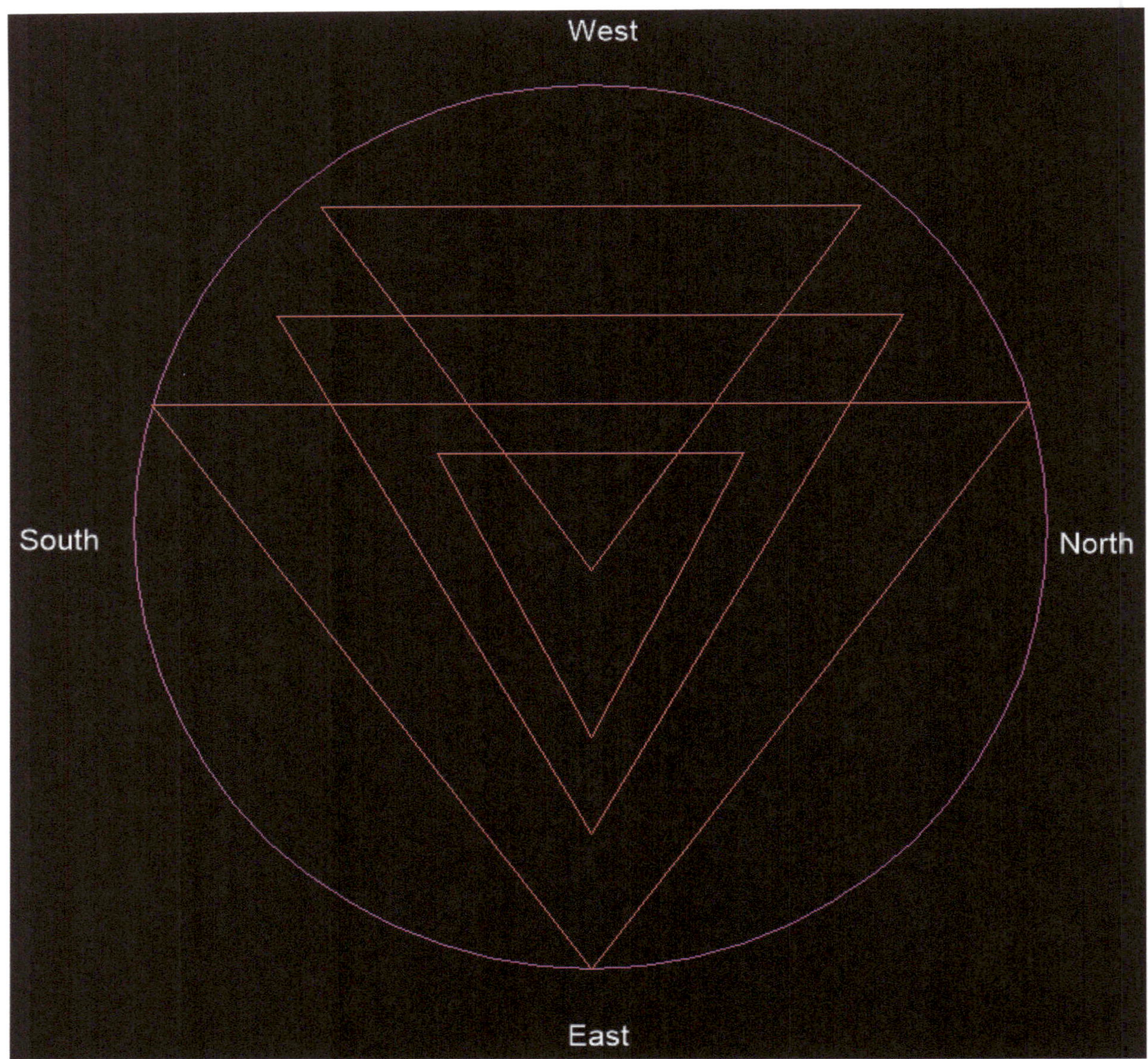

Figure 68 **Four Downward Apex**

One Downward Apex Union Triangle with Bindu

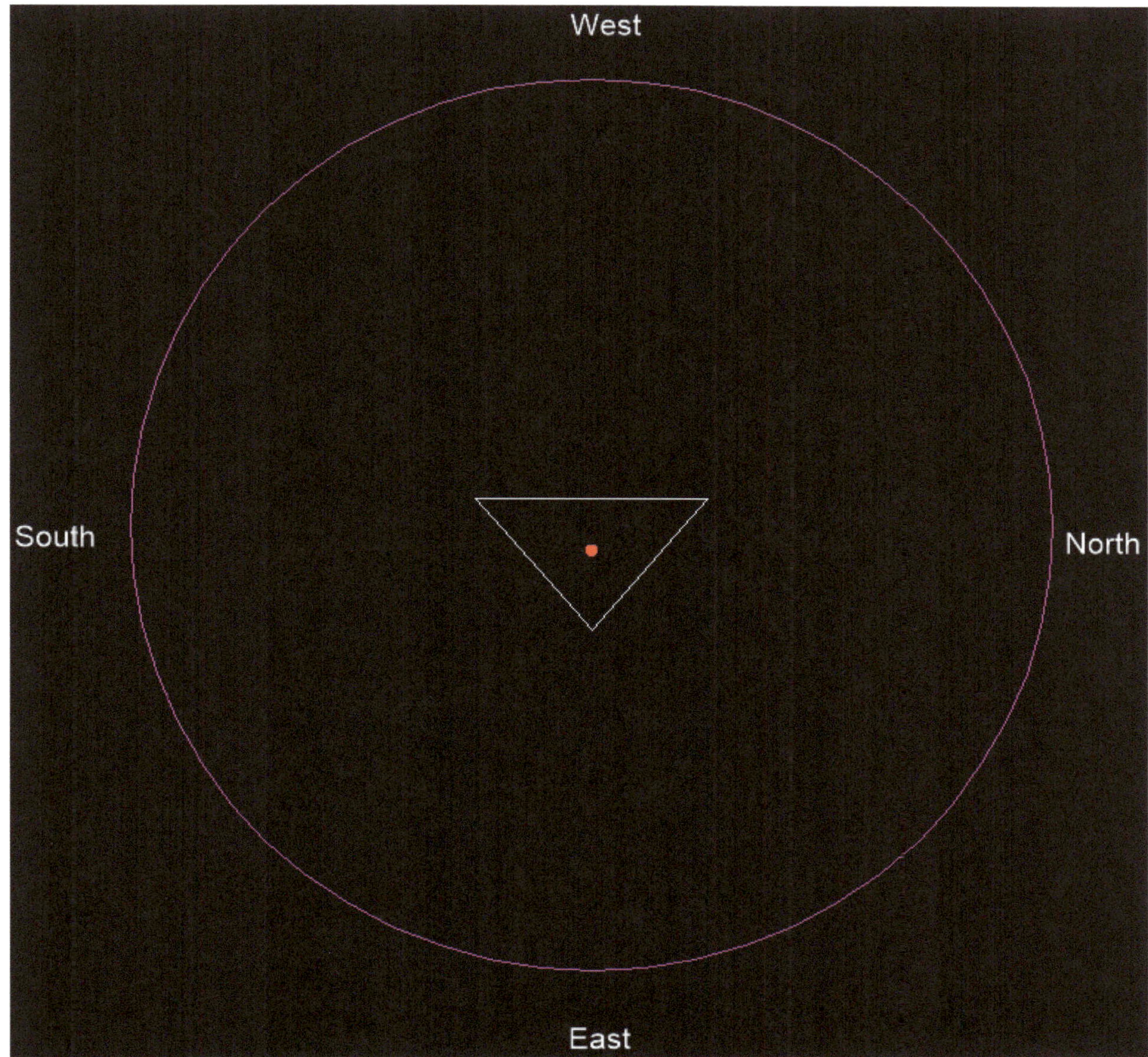

Figure 69 **Innermost Downward Apex with Bindu**

Sri Yantra with Sanskrit Verses and Seed Sounds

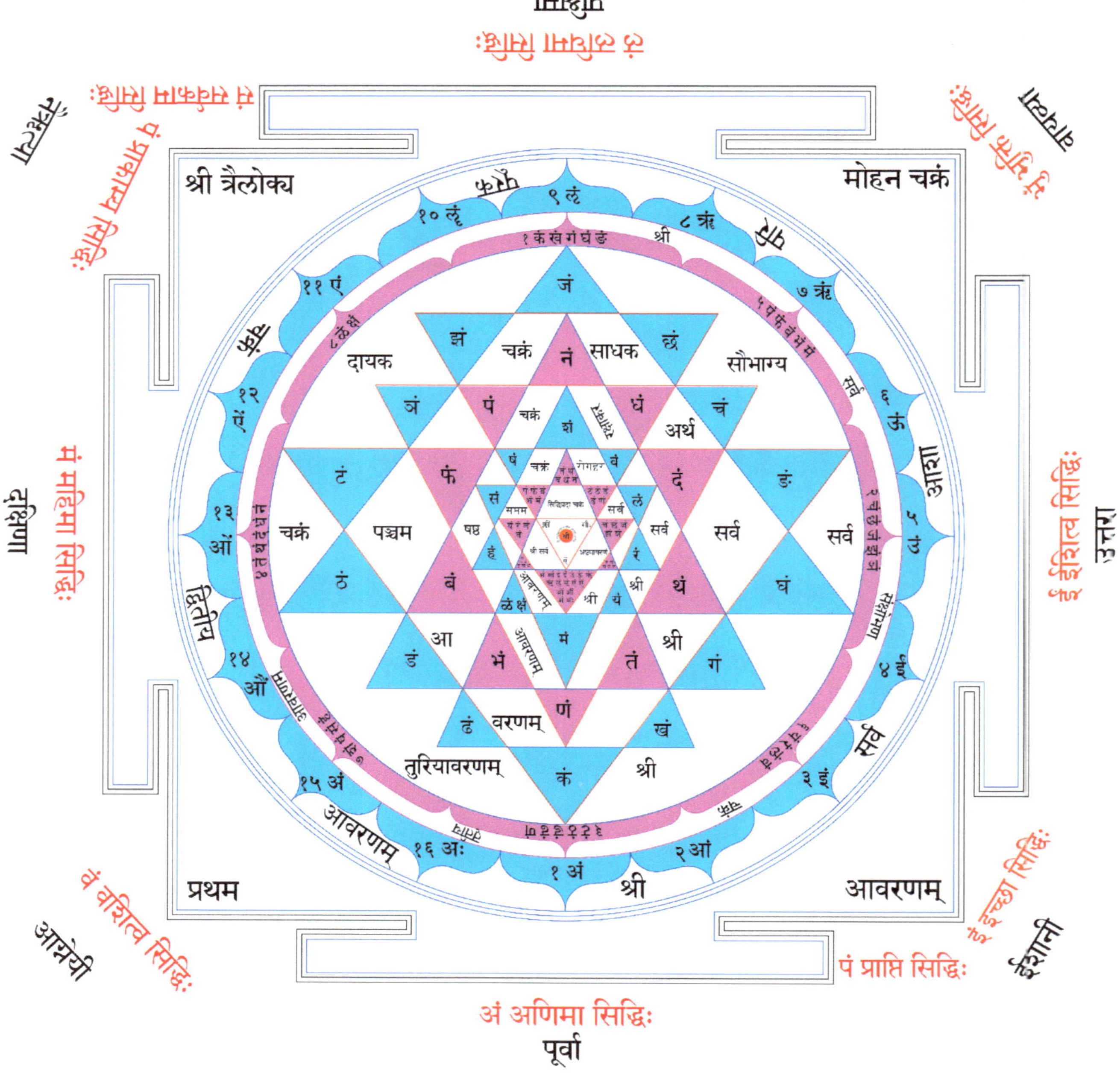

श्री यन्त्रम्

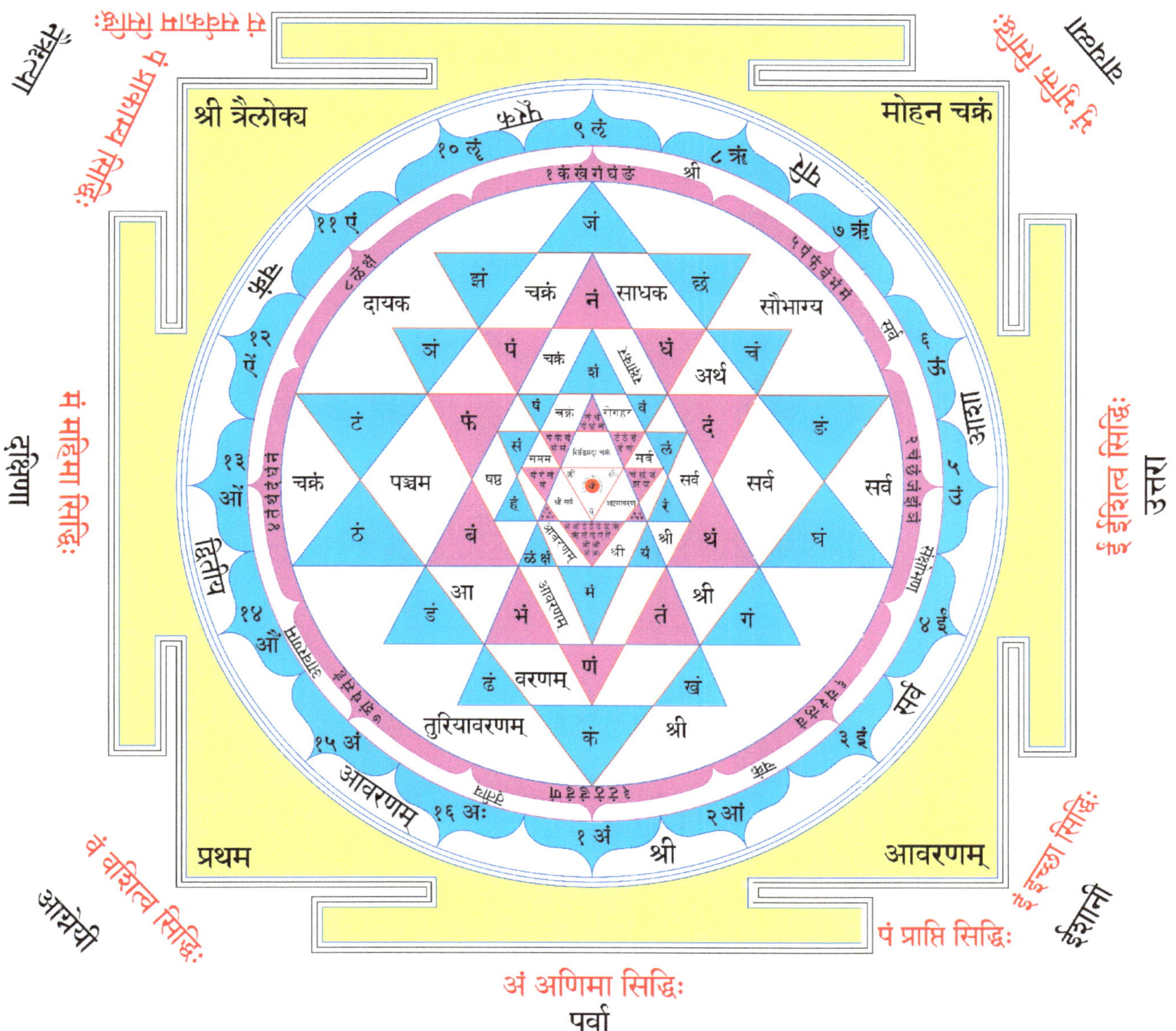

|| ॐ श्री ललिता महात्रिपुरसुन्दर्यै नमः ||

श्री यन्त्रम्

Sandhi Points 24 Dual Intersections

The Sri Yantra consists of **dual** intersecting points that are 24 in number. These junctions are known as Sandhi (Joint), where two lines cross each other.

Sri Yantra

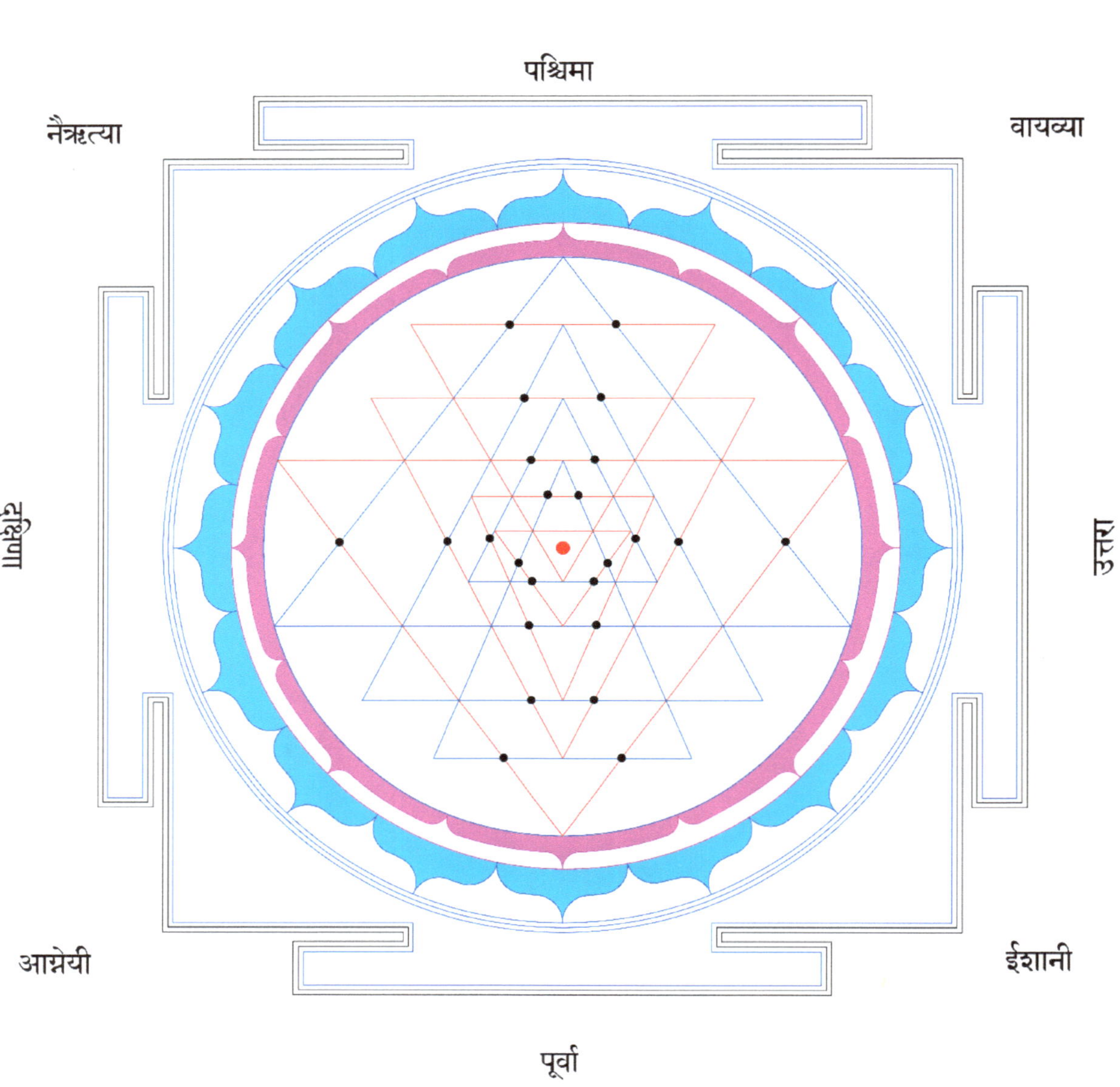

The 24 Sandhi Junctions in the Sri Yantra, shown by dots.

Marma Points 18 Triple Intersections

The Sri Yantra consists of **triple** intersecting points that are 18 in number. These junctions are known as Marma (presence of the divinity), where three lines cross each other.

Sri Yantra

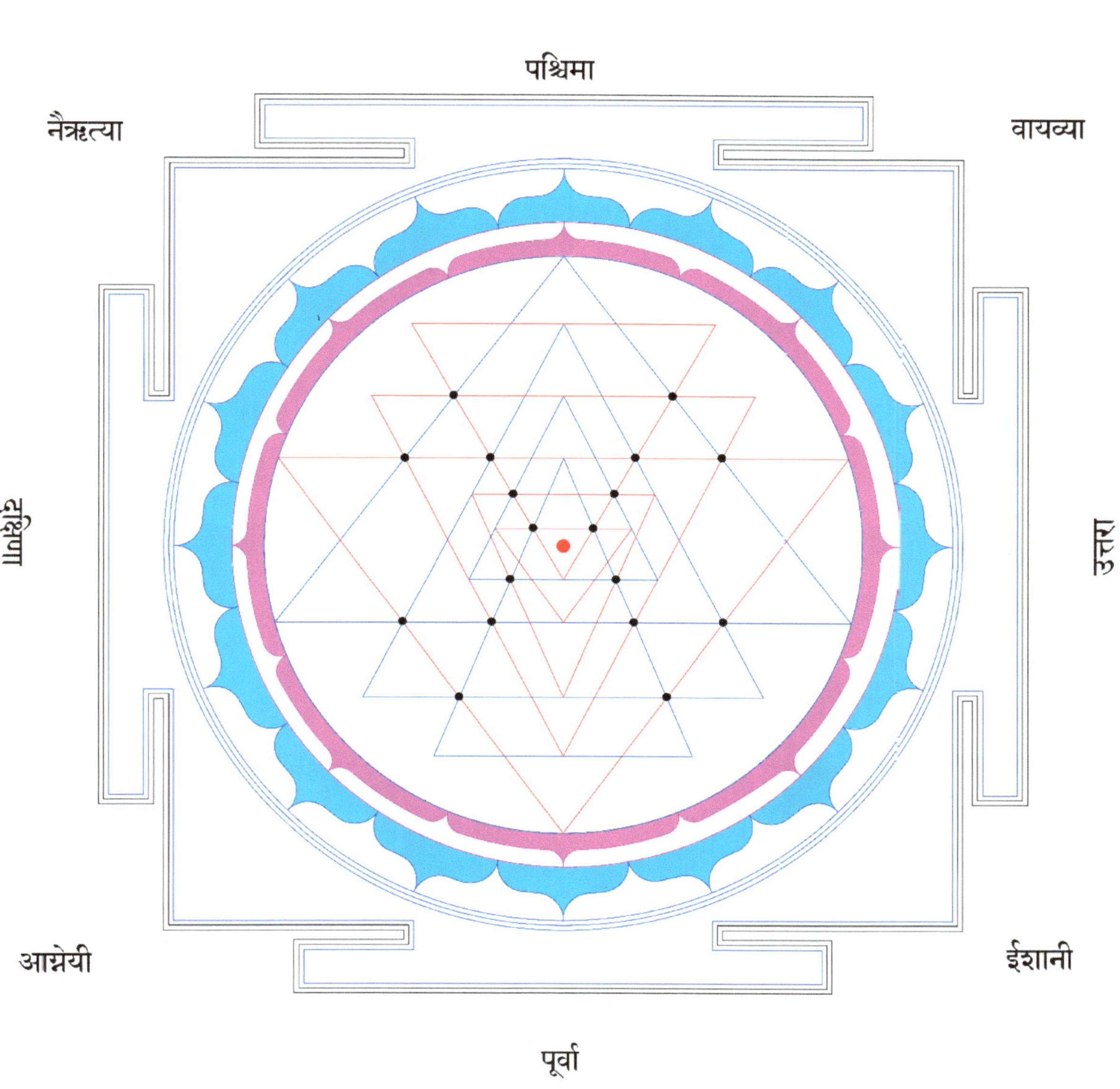

The 18 Marma Junctions in the Sri Yantra, shown by dots.

Probable errors in our Sri Yantra construction

- Innermost Triangle T5d is not completely Equilateral
- Bindu and Center of Circle circumscribing T1u and T1d are displaced slightly
- Choice of 7 as multiplier for 3 4 5 Triangle T1d may be adjusted for a better diagram

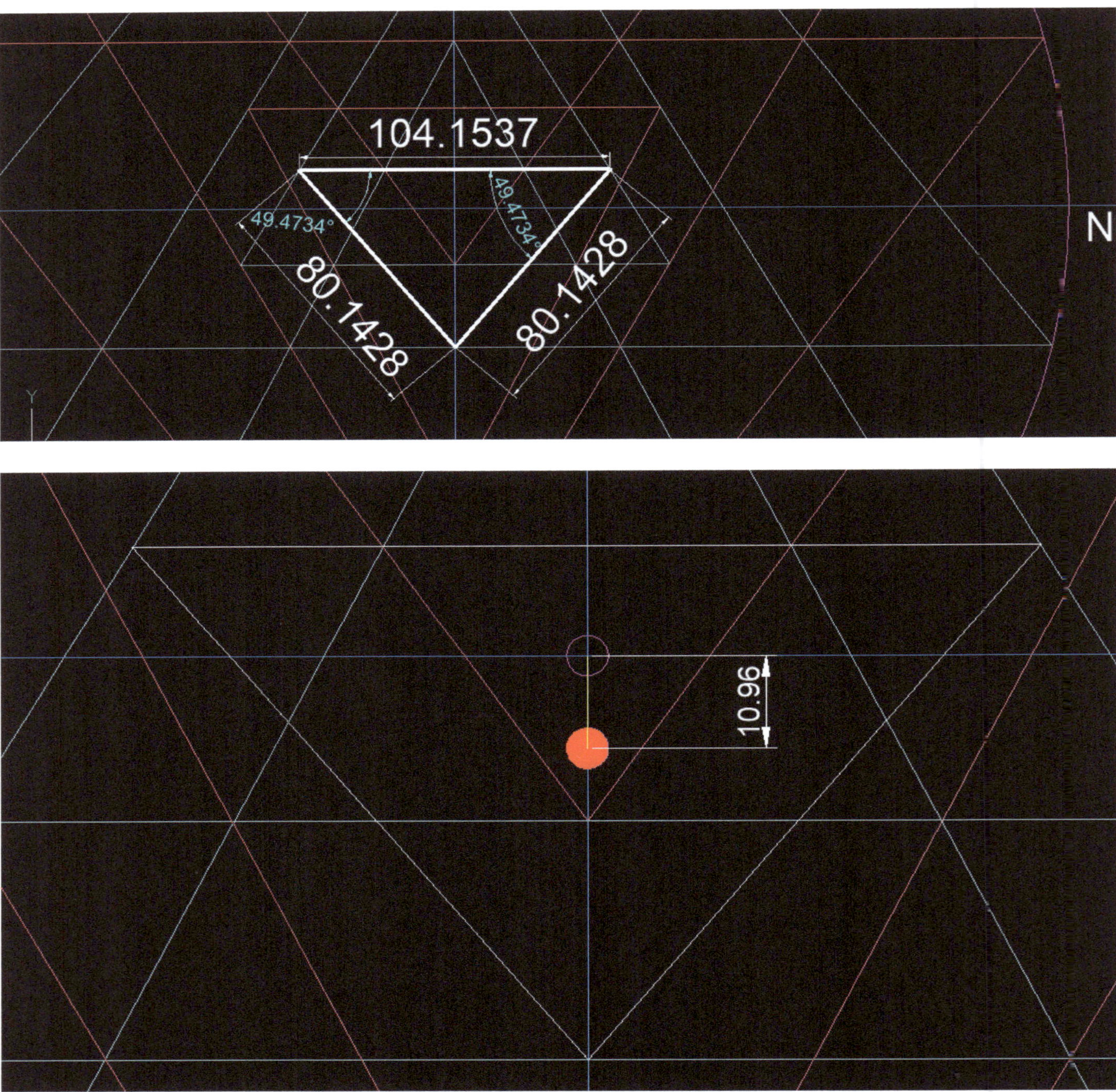

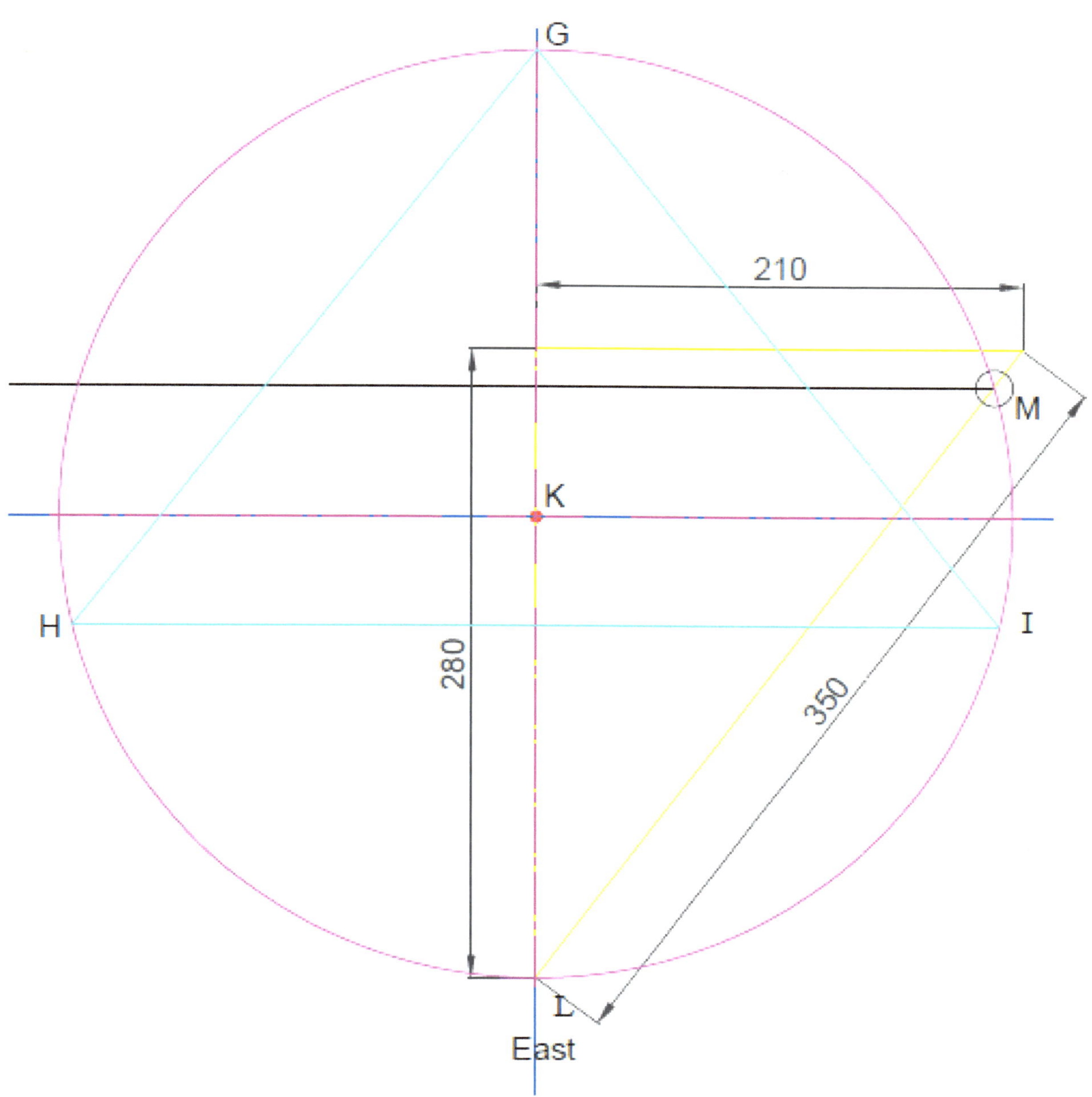

G
210
M
K
H
I
280
350
L
East

Golden Ratio Triangle in Sri Yantra

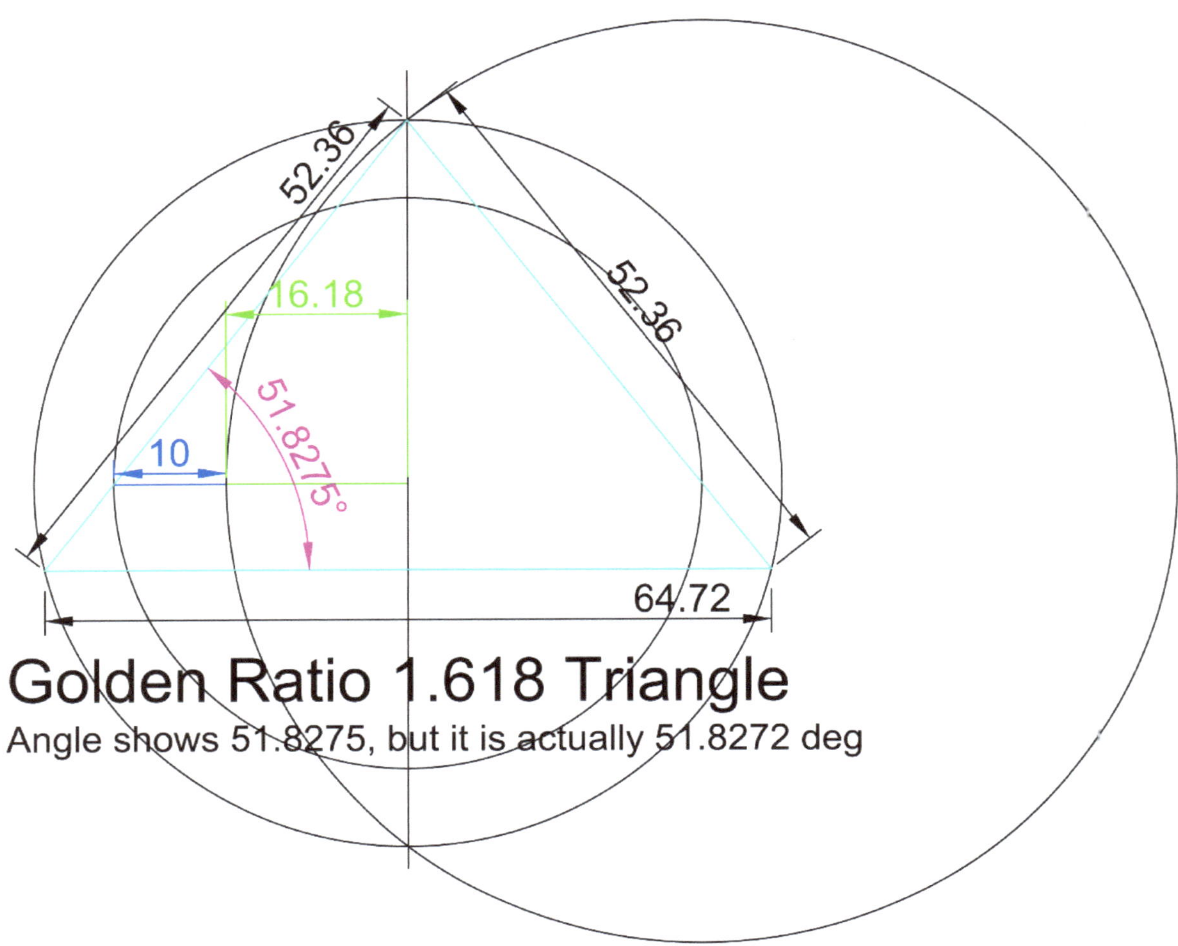

The first biggest Upward Apex Triangle in the Sri Yantra is built using the Golden Ratio found in nature.

What is the Golden Ratio? First let us see the Fibonacci series.

The Fibonacci numbers are **integers** in the following sequence: **0, 1, 1, 2, 3, 5, 8, 13, 21, 34, 55, 89, 144**..i.e., a series of numbers in which the next number is made by adding the previous two integers

Now let us see the ratio of two successive Fibonacci numbers.

i.	1/0	invalid
ii.	1/1 = 1	
iii.	2/1 = 2	
iv.	3/2 = 1.5	
v.	5/3 = 1.6666	
vi.	8/5 = 1.6	
vii.	13/8 = 1.625	
viii.	21/13 = 1.6153	
ix.	34/21 = 1.6190	
x.	55/34 = 1.6176	

xi.   89/55 = 1.6181

xii.   144/89 = 1.6179

xiii.   233/144 = 1.6180

xiv.   377/233 = 1.6180

xv.   610/377 = 1.6180

After a while this ratio series becomes more or less stable on the irrational number 1.6180, and this number is known as the Golden Ratio.

It is also called the golden section, golden mean, or divine proportion. In mathematics, the irrational number (1 + Square root of 5)/2, often denoted by the Greek letter Phi ϕ, which is approximately equal to 1.618. Thus Golden Ratio = 1.6180 = $(1 + \sqrt{5})/2 = \phi$

Now consider the inverse $1/\phi = 0.6180 = \phi -1$ which is called phi and denoted by φ (lowercase).

a + b = 0.618 + 1 = 1.6180 = Phi = ϕ = Uppercase.

0.6180 = phi = φ = Lowercase.

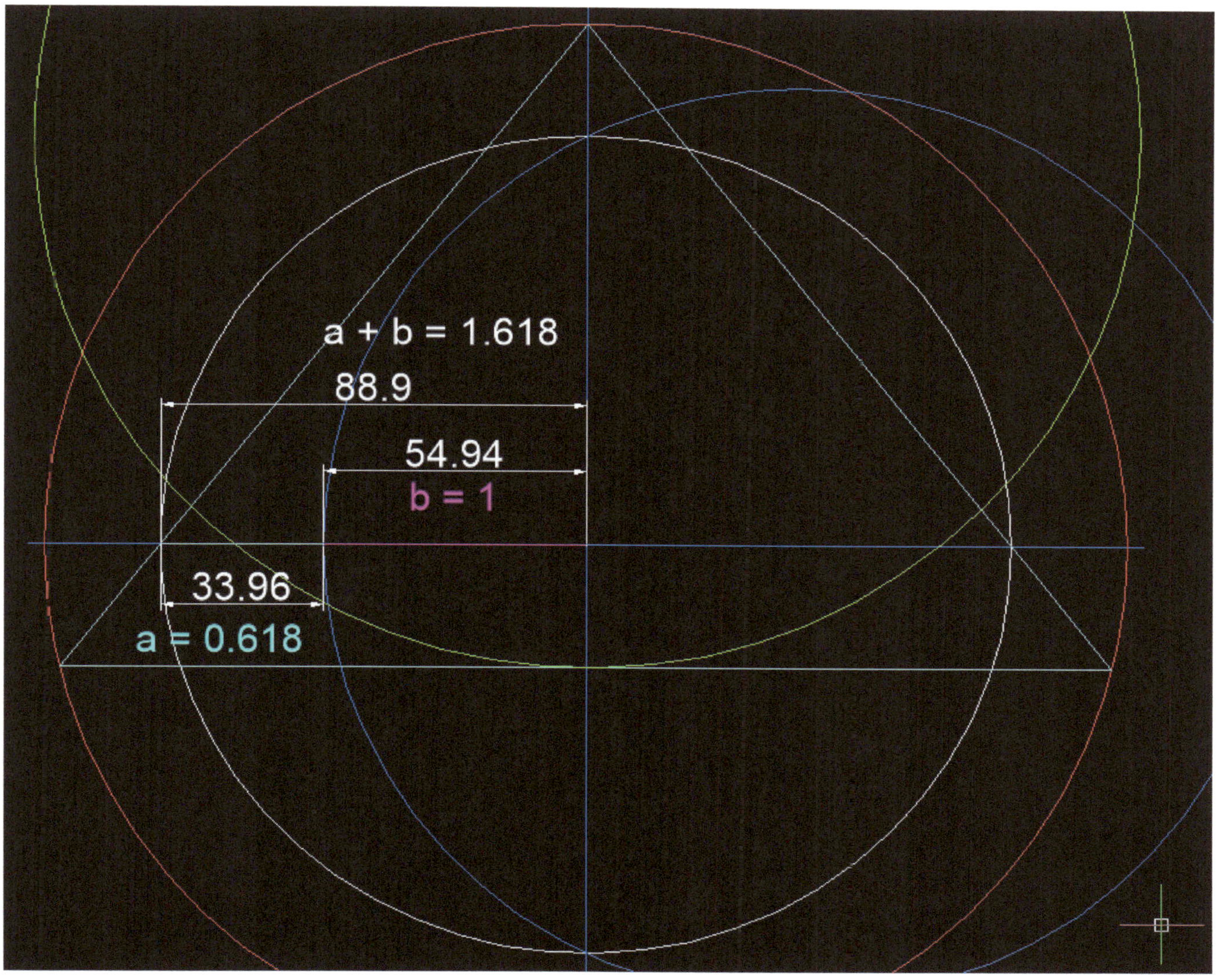

Sri Yantra is from Bhavana Upanishad भावनोपनिषद्

The Bhavanopanishad is attributed to the Atharvaveda. It expounds the innermost intense feelings of the Soul, its basic desire, and ways and means to connect and unite with the ultimate consciousness.

ॐ भद्रं कर्णेभिः श्रृणुयाम देवाः । भद्रं पश्येम माक्षभिर् यजत्राः । स्थिरैरङ्गैस् तुष्टुवां सस्तनूभिः । व्यशेम देवहितं यदायुः ॥

स्वस्ति न इन्द्रो वृद्धश्रवाः । स्वस्ति नः पूषा विश्ववेदाः । स्वस्ति नस्ताक्ष्र्यो अरिष्टनेमिः । स्वस्ति नो बृहस्पतिर्दधातु ॥

ॐ शान्तिः शान्तिः शान्तिः ॥

ॐ आत्मानम् अखण्ड-मण्डलाकारम् आवृत्य सकल-ब्रह्माण्ड-मण्डलं स्वप्रकाशं ध्यायेत् । श्री गुरुः सर्वकारणभूता शक्तिः ॥ १ ॥ तेन नव-रन्ध्र-रूपो देहः ॥ २ ॥ नव-शक्ति-रूपश्च श्री-चक्रम् ॥ ३ ॥ वाराही पितृरूपा । कुरुकुल्ला बलिदेवता माता ॥ ४ ॥ पुरुषार्थाः सागराः ॥ ५ ॥ देहो नव-रत्न-द्वीपः ॥ ६ ॥ त्वक्-आदि-सप्त-धातुभिः अनेकैः संयुक्ताः सङ्कल्पाः कल्पतरवः ॥ ७ ॥ तेजः कल्पक-उद्यानम् ॥ ८ ॥ रसनया भाव्यमाना मधुः-आह्ल-तिक्त-कटु-कषाय-लवण-रसाः षड् ऋतवः । क्रिया-शक्तिः पीठम् । कुण्डलिनी ज्ञान-शक्तिः गृहम् । इच्छा-शक्तिः महात्रिपुरसुन्दरी ॥ ९ ॥ ज्ञाता होता ज्ञानम् अग्निः ज्ञेयश्च हविः । ज्ञातृ-ज्ञान-ज्ञेयानाम् अभेदभावनश्च श्री-चक्र-पूजनम् ॥ १० ॥

Enclosures

1st. नियति सहिताः शुद्धाः आद्यो नव-रसाः अणिमा आदयः । काम-क्रोध-लोभ-मोह-मद-मात्सर्य-पुण्य-पापमय्यी ब्राह्मि आदि अष्ट शक्तयः ॥ ११ ॥ आधर-नवकं मुद्रा-शक्तयः ॥ १२ ॥

2nd. पृथिवी-अप-तेजः-वायु-आकाश-श्रोत्र-त्वक्-चक्षुः-जिह्वा-घ्राण-वाक्-पाणि-पाद-पायु-उपस्थ-मनोविकाराः कामाकर्षिणी आदि षोडश शक्तयः ॥ १३ ॥

3rd. वचन-आदान-गमन-विसर्ग-आनन्द-हान-उपादान-उपेक्षा-बुद्धयः अन्-अङ्ग-कुसुमा आदि शक्तयः अष्टौ ॥ १४ ॥

4th. अलम्बुषा कुहूः विश्वोदरा वारणा हस्तिजिह्वा यशोवती पयस्विनी गान्धारी पूषा शङ्खिनी सरस्वती इडा पिङ्गला सुषुम्ना चेति चतुर्दशः नाड्यः । सर्व-संक्षोभिणी आदि चतुर्दशारगा देवताः ॥ १५ ॥

5th. प्राण-अपान-व्यान-उदान-समान-नाग-कूर्म-कृकर-देवदत्त-धनञ्जयाः इति दशवायवः । सर्व-सिद्धिप्रदा देव्यो बहिर्दशारगा देवताः ॥ १६

6th. एतद् वायु-दशक संसर्गोपाधिभेधेन रेचक-पाचक-शोषक-दाहक-प्लावकाःअमृतमिति प्राणमुख्यत्वेन पञ्चधा जठर-अग्निः भवति । १७ क्षारकः उद्धारकः क्षोभकः मोहकः जृम्भकः इति नाग-प्राधान्येन पञ्चविधोऽस्ति । तेन मनुष्याणां देहानां भक्ष्य-भोज्य-चोष्य-लेह्य-पेयात्मकं पञ्चविधम्-अन्नं पाचयन्ति ॥ १८ ॥ एता दश वह्निकलाः सर्वज्ञाद्याः अन्तर्दशारगा देवताः ॥ १९ ॥

7th. शीत-उष्ण-सुख-दुःख-इच्छाः सत्त्व-रजस्-तमोगुणाः वशिनी आदि शक्तयः अष्टौ ॥ २० ॥

8th. शब्द-स्पर्श-रूप-रस-गन्धाः पञ्च-तन्मात्राः पञ्च-पुष्पबाणाः ॥ २१ ॥ मनः इक्षु-धनुः ॥ २२ ॥ रागः पाशः ॥ २३ ॥ द्वेषः अङ्कुशः ॥ २४ अव्यक्त-महत्तत्त्वम्-महतअहङ्कारः इति कामेश्वरी-वज्रेश्वरी-भगमालिनी अन्तस् त्रिकोण-अग्रगा देवताः ॥ २५ ॥

9th. निरुपाधिका संविदेव कामेश्वरः ॥ २६ ॥ सदानन्दपूर्णा स्वात्मैव परदेवता ललिता ॥ २७ ॥ लौहित्यमेतस्य सर्वस्य विमर्शः ॥ २८ ॥ अनन्यचित्तत्वेन च सिद्धिः ॥ २९ ॥ भावनायाः क्रियाः उपचाराः ॥ ३० ॥ अहं त्वम् अस्ति नास्ति कर्तव्यम् अकर्तव्यम् उपासितव्यम् इति विकल्पानाम् आत्मनि विलापनं होमः ॥ ३१ ॥ भावना विषयाणाम् अभेद-भावना तर्पणम् ॥ ३२ ॥ पञ्चदश तिथि-रूपेण कालस्य परिणामा अवलोकनं पञ्चदश नित्याः ॥ ३३ ॥ एवं मुहूर्त-त्रितयं मुहूर्त-द्वितयं मुहूर्त-मात्रं वा भावनापरो जीवन्मुक्तो भवति । स एव शिवयोगी इति कथ्यते ॥ ३४ ॥ कादिमतेन अन्तश्चक्र-भावनाः प्रतिपादिताः ॥ ३५ ॥ य एवं वेद । सः अथर्वशिरः अधीते ॥ ३६ ॥ इत्युपनिषत् ॥

ॐ भद्रं कर्णेभिः श्रृणुयाम देवाः । भद्रं पश्येम माक्षभिर् यजत्राः । स्थिरैरङ्गैस् तुष्टुवां सस्तनूभिः । व्यशेम देवहितं यदायुः ॥ स्वस्ति न इन्द्रो वृद्धश्रवाः । स्वस्ति नः पूषा विश्ववेदाः । स्वस्ति नस्ताक्ष्र्यो अरिष्टनेमिः । स्वस्ति नो बृहस्पतिर्दधातु ॥ ॐ शान्तिः शान्तिः शान्तिः ॥

Avaranas - the nine enclosures, and their Sanskrit Alphabet Letters, i.e., Seed Sounds.
नव आवरण A discussion and drawing of the Nine Enclosures.

1. त्रैलोक्यमोहन चक्रं that which attracts, impresses and infatuates all the three worlds

2. सर्व आशापरिपूरक चक्रं that which fulfills desires and needs of all

3. सर्व संक्षोभण चक्रं that which attracts, impresses and infatuates all the three worlds

4. सर्व सौभाग्यदायक चक्रं that which proves to be fortunate for all

5. सर्वार्थसाधक चक्रं that which provides for and nourishes all

6. सर्व रक्षाकर चक्रं that which protects and guards all

7. सर्व रोगहर चक्रं that which banishes illness and grief from all

8. सर्व सिद्धिप्रद चक्रं that which enhances the talents of each and blossoms everyone

9. सर्व आनन्दमय चक्रं that which infuses bliss, the pure undiluted joy in everyone

Directions in a Sri Yantra
In the Sri Yantra, we travel **anticlockwise** since East is depicted below. Sequence is:
i.East ii.NorthEast iii.North iv.NorthWest v.West vi.SouthWest vii.South viii.SouthEast

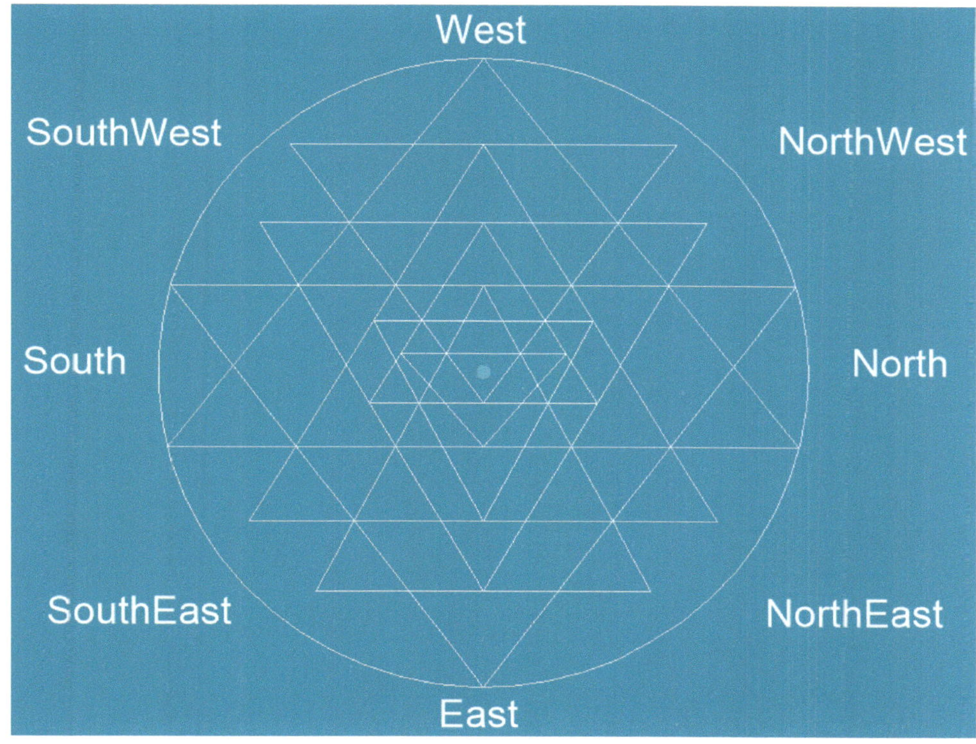

In Sanskrit we use the **feminine** spellings for the directions and the divinities.
पूर्वा, ईशानी, उत्तरा, वायव्या, पश्चिमा, नैर्ऋत्या, दक्षिणा, आग्नेयी ।

पूर्वा East, ईशानी NorthEast, उत्तरा North, वायव्या NorthWest, पश्चिमा West, नैर्ऋत्या SouthWest, दक्षिणा South, आग्नेयी SouthEast.

Legend

The 9 Basic Triangles of Sri Yantra		
SN	Acronym	Remarks
1	T1u	1st Triangle. Triangle 1st Upward Apex. It is an isosceles Golden Ratio Triangle
2	T1d	2nd Triangle. Triangle 1st Downward Apex. Largest triangle of all.
3	T2u	3rd Triangle. Triangle 2nd Upward Apex
4	T2d	4th Triangle. Triangle 2nd Downward Apex
5	T3u	5th Triangle. Triangle 3rd Upward Apex
6	T3d	6th Triangle. Triangle 3rd Downward Apex
7	T4u	7th Triangle. Triangle 4th Upward Apex
8	T4d	8th Triangle. Triangle 4th Downward Apex
9	T5d	9th Triangle. Triangle 5th Downward Apex. Innermost smallest Triangle with B ndu

Acronyms are arbitrary names used for clarity in this book. The triangle numbering for Upward Apex Triangles is simply done based on **top to bottom** vertical sequence. Similarly for Downward Apex Triangles the mirror sequence **bottom to top** is used. Except for Innermost triangle.

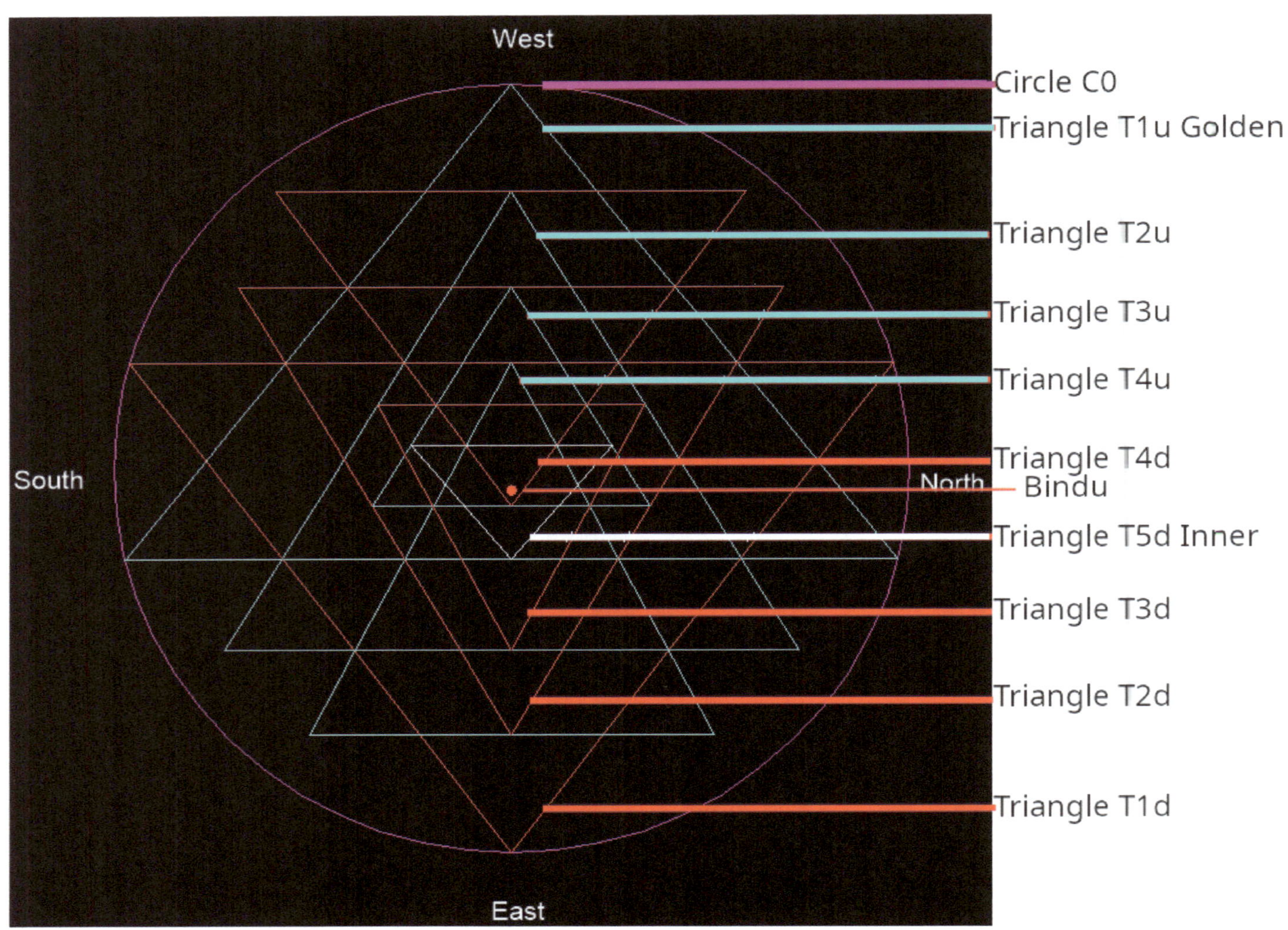

Suggested Colors

Upward Apex Shiva Triangles (Purusha). **The Being**		
T1u	Cyan (Blue)	Suggested color is Blue, however cyan is used for contrast in print and screen
T2u	Cyan (Blue)	
T3u	Cyan (Blue)	
T4u	Cyan (Blue)	
Downward Apex Shakti Triangles (Prakriti). **The Ambience**		
T1d	Red	
T2d	Red	
T3d	Red	
T4d	Red	
Innermost smallest Triangle with Bindu. **The Union**. Yoga		
T5d	White	Bindu is the Red Dot inside

Sequence of Triangle Completion while Drawing

The 9 Basic Triangles of Sri Yantra with Bindu and enclosed Circle, and Directions		
Sequence	Acronym	Remarks
1	1. T1u	Triangle 1st Upward Apex. It is an isosceles Golden Ratio Triangle
1a	Circle C0	Circle that encircles and touches T1u, T1d. Add circle center Cross Hair.
2	3 4 5 Triangle	3 4 5 Right Angle Triangle with multiplier 7, i.e. 210 280 350 Triangle.
2a	2. T1d	Triangle 1st Downward Apex. Slightly larger than T1u.
3	5. T4u	Triangle 4th Upward Apex
4	3. T2u	Triangle 2nd Upward Apex
5	8. T4d	Triangle 4th Downward Apex
6	6. T2d	Triangle 2nd Downward Apex
7	7. T3u	Triangle 3rd Upward Apex
8	4. T3d	Triangle 3rd Downward Apex
9	9. T5d	Triangle 5th Downward Apex. Innermost smallest Triangle with Bindu
9a	Bindu	Bindu Red Dot
10	ESWN	Cardinal Directions. We switch off Cross Hair and get Sri Yantra Core

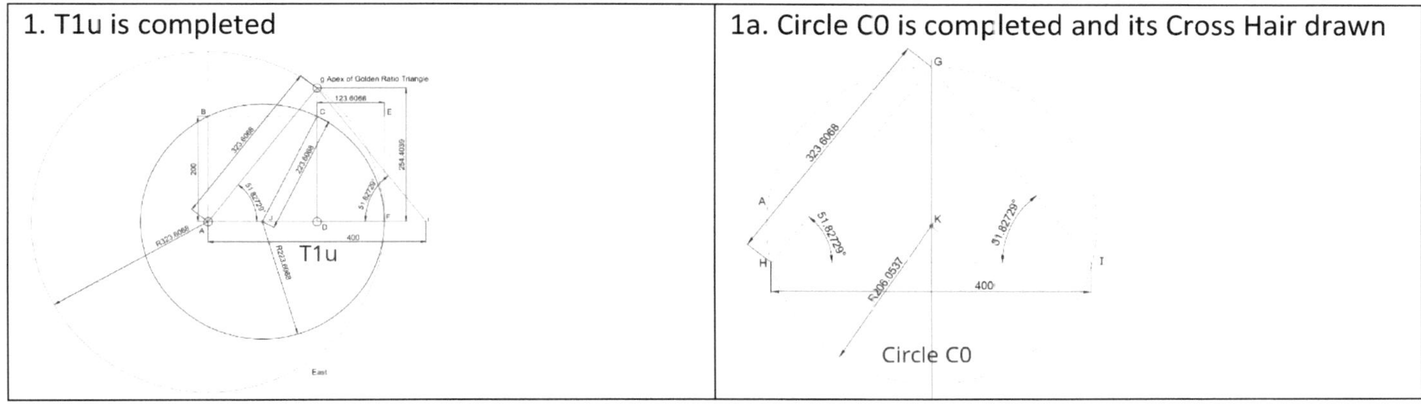

1. T1u is completed

1a. Circle C0 is completed and its Cross Hair drawn

<table>
<tr><td>

2. 3 4 5 Right Angle Triangle is drawn

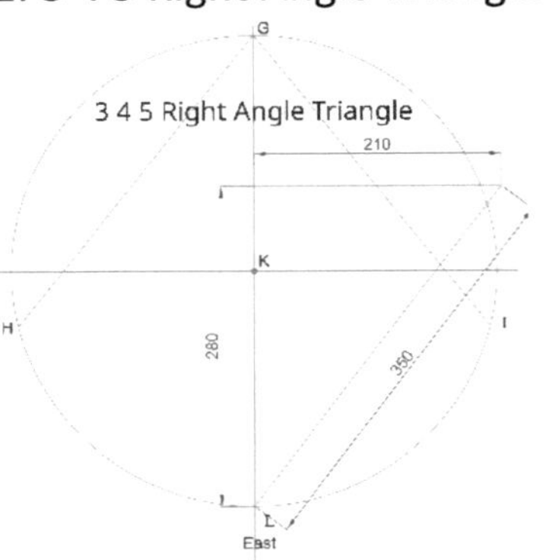

</td><td>

2a. T1d is completed

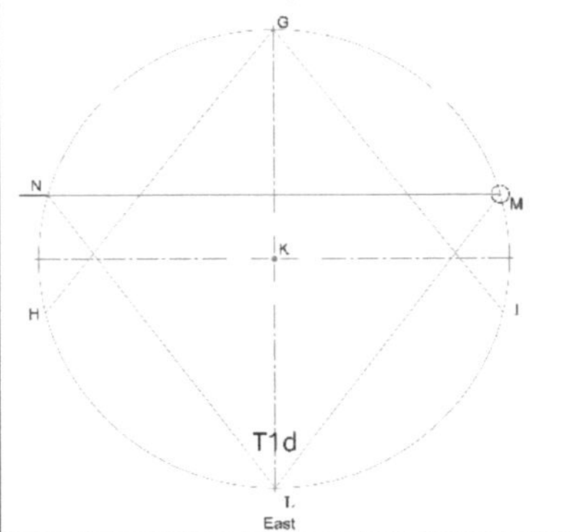

</td></tr>
</table>

3. T4u is completed

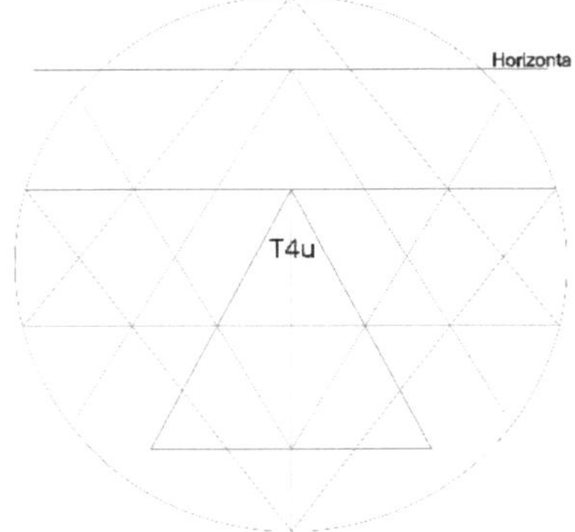

4. T2u is completed

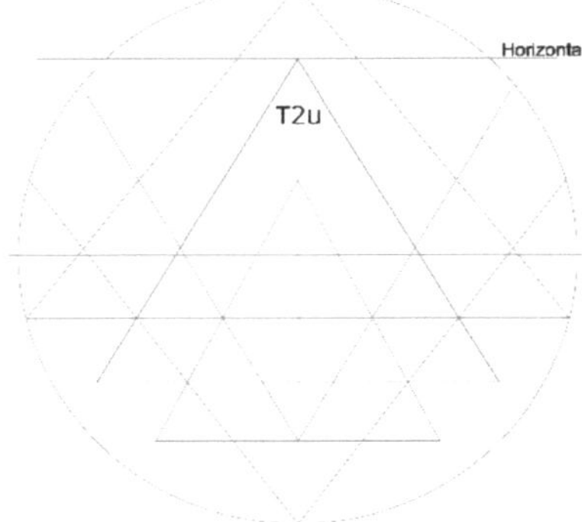

5. T4d is completed

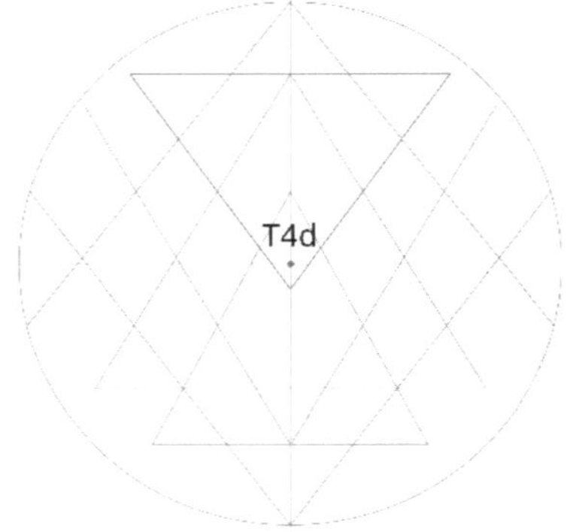

6. T2d is completed

7. T3u is completed

8. T3d is completed

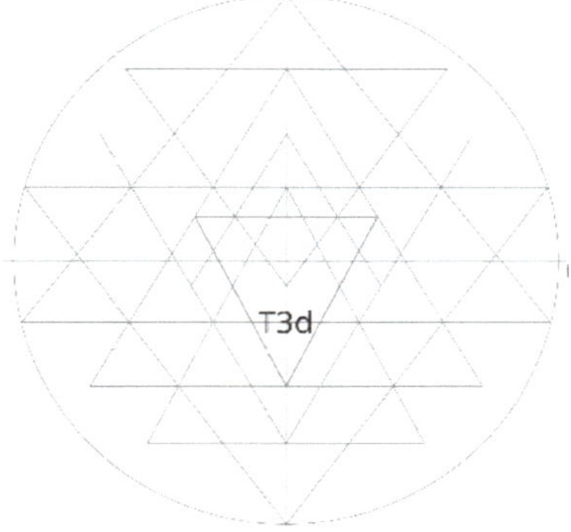

9. T5d is completed	9a. Bindu is added incenter of T5d

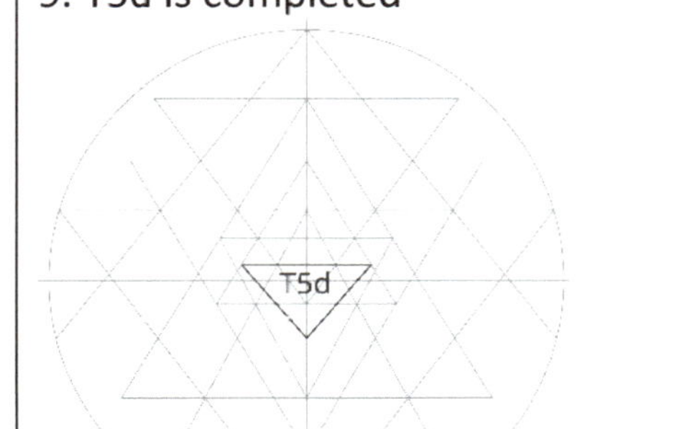

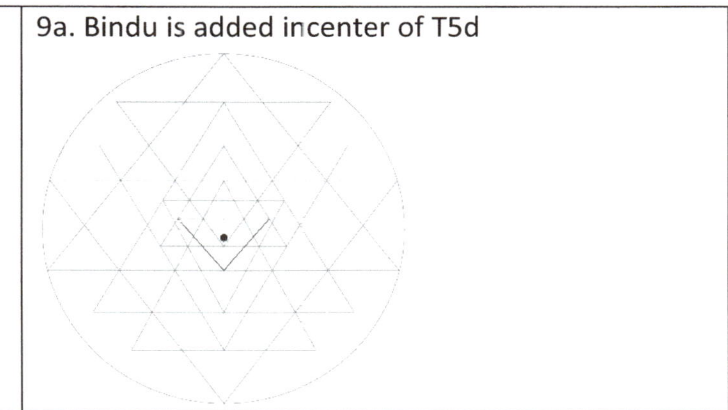

10. ESWN Cardinal Directions and Cross Hair off gives Sri Yantra Core

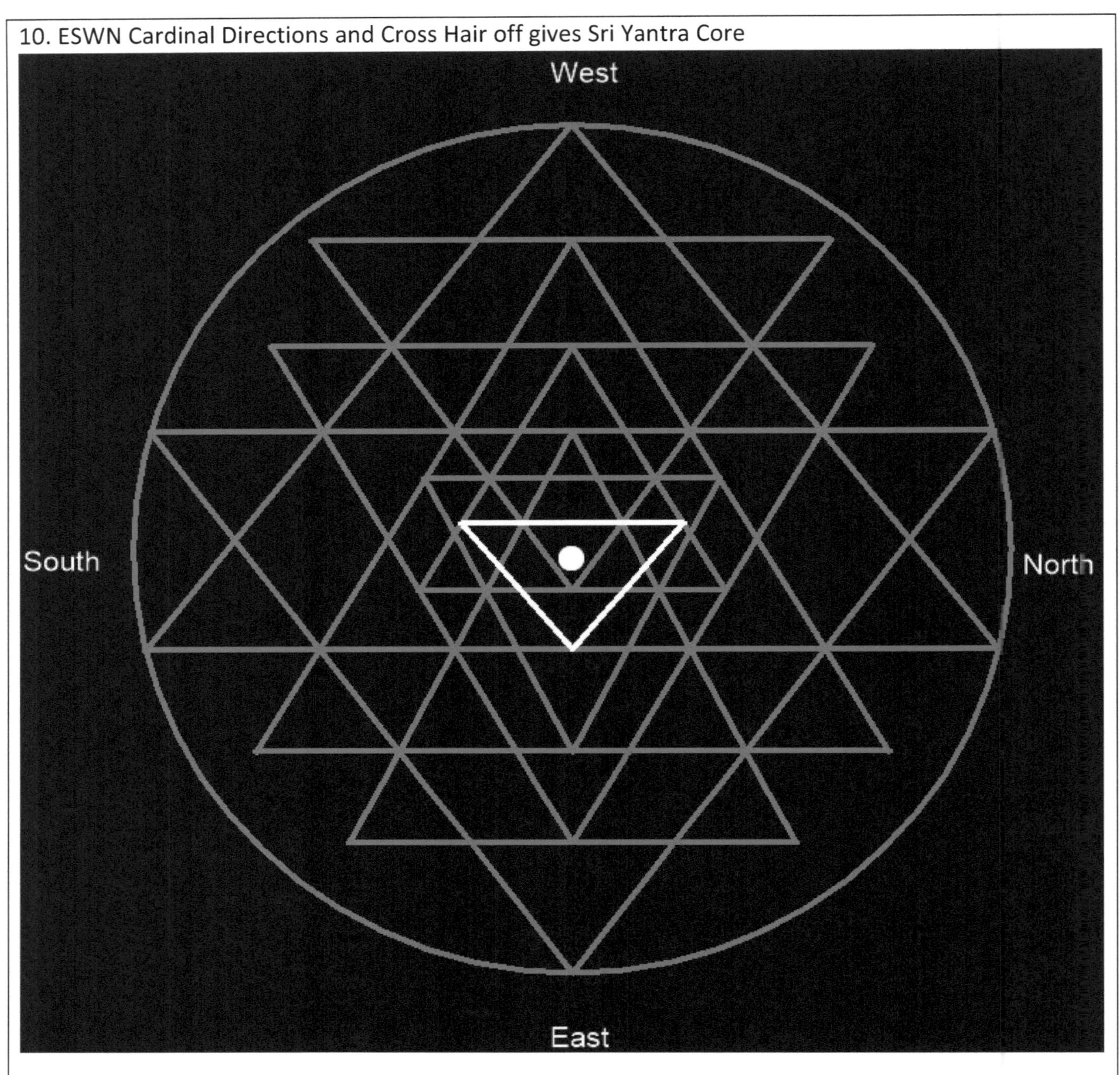

Sequence of Construction

1. Begin line 100mm. Starting point A, Ending point J. Line AJ = 100mm.
2. Extend to Square of each side 200mm. Square = ABCD.
      2a. Check dimensions with precision of 4 decimal places
3. Mark Halfbase point of Square at 100, join points for hypotenuse of right angle triangle inside Square
4 Draw a Circle C1 with center at J Halfbase mark of Square, and radius = hypotenuse = 223.6068
      4a. Recheck dimensions with precision of 4 decimal places
5. Extend Square to make Golden Ratio Rectangle ABEF, touching circumference of Circle C1 at point F.
      5a. Check that dimensions of Rectangle are, longer side 323.6068, shorter side 200
6. Draw Circle C2 with center at start point A, radius = longer side of rectangle = 323.6068
7. Extend side DC of Square vertically upwards to touch Circle C2. Name this new point g.
      7a. To double check, extend side CD of Square vertically downwards to touch Circle C2.
      7b. Check that lengths of the extended segments are equal.
8. Join A and g to get the side Ag for Golden Ratio Triangle T1u. Ag = 323.6068
      8a. Copy and Mirror Ag to get the other side gi of Isosceles Triangle T1u.
      8b. Join Ai to get the base of Golden Ratio Triangle T1u = Agi.
      8c. Dimensions of T1u, isosceles sides = 323.6068 each, base = 400, isosceles angles = 51.82729°.

We need to draw a circle that **circumscribes** Triangle T1u. Such a circle is found by bisecting each **side** of the triangle at 90°, and where these bisectors meet that is the center of the circle to circumscribe.

9. Bisect sides Ag and Ai of T1u to intersect at a new point O that becomes the center of new Circle C0.
10. We can also draw a circle using 3-point, then select points A, G, I.
11. Using point O as center, draw Circle C0 that circumscribes Triangle T1u.
12. Check that Radius of Circle C0 so made = 206.0537mm
13. Draw centermark of circle C0 and extend it to make Cross Hair, Horizontal, Vertical with center at O.

Now we do the most important part. This is a variable, so choosing a different value shall give a different construction of the Sri Yantra. We choose a multiplier for the 3 4 5 Right Angle Triangle. Criteria is that **base should be bigger** than radius of C0. Also, **area** of next triangle T1d should be **slightly larger** than T1u. Let us keep the multiplier as 70, then 3:4:5 = 70x3 : 70x4 : 70x5 = 210:280:350. Check that $a^2 + b^2 = c^2$, $210^2 + 280^2 = 350^2$, $44100 + 78400 = 122500$. Thus Right Angle Triangle horizontal side = 210, vertical side = 280, hypotenuse = 350.

14. Draw the vertical side = 280 of the Right Angle Triangle from East, where cross hair vertical touches the Circle C0. At its upper point, draw a horizontal of length 210. Join the two points to complete the hypotenuse.
15. Now we need to chop off the hypotenuse of this Right Angle Triangle at the point where it touches the Circle C0. And redraw the horizontal from this point to go all the way from North to South, to touch the periphery of Circle C0. Using this chord as the base for our next Triangle T1d, draw its sides to touch the Circle C0 at East.
16. Complete T1d. It is **slightly larger** than T1u in **area**.
17. Check Dimensions of T1d.
18. Add the Cross Hair by using centermark of Circle C0.

19. Get the HalfBase length of T1d = 197.5844, flip it vertically, and align its lower point where the cross hair vertical intersects base of T1u.

20. This gives us the Horizontal guide for T2u Apex and T4d base.

21. Using the new temporary Sandhi points for T2u sides and its Apex, we Draw the T2u sides guides.

22. Similarly, we get the halfbase length of T1u, flip it vertical and align its upper point where the cross hair vertical intersects base of T1d.

23. This gives us the Horizontal guide for T2d Apex and T4u base.

24. Using the new temporary Sandhi points for T2d sides and its Apex, we Draw the T2d sides guides.

25. Using the new temporary Sandhi points for T4u sides and its Apex, we Draw the T4u Triangle completely, since its base guide is already there.

26. Using the new temporary Sandhi points for T2u base, we Draw the T2u base guide.

27. Since T2u Apex is already known, and its sides guides are already drawn, we complete T2u.

28. Check dimensions of T2u.

29. We mark four new temporary Sandhi points for finding T4d Apex. Join these two segments. The new temporary sandhi points so made shall give the horizontal guide for T4d Apex and also T3u base.

30. Draw sides for T4d.

31. Draw base for T4d.

32. Complete T4d.

33. Check Dimensions T4d.

34. We get the new sandhi points for T2d baseline.

35. Draw the guides for T2d base, and since Apex and sides are already present, we complete T2d.

36. Check T2d dimensions are isosceles.

37. We get the new sandhi point for T3u Apex on the base of T2d.

38. Since base and sides guides are already drawn, we complete T3u.

39. Check T3u dimensions are isosceles.

40. Now we obtain new sandhi points for T3d base guide and draw T3d base guideline.

41. We also get Sandhi point for T3d Apex, so complete T3d.

42. Check T3d dimensions are isosceles.

43. We now obtain sandhi point for T5d Apex, and T5d baseline.

44. Complete T5d.

45. Check T5d dimensions are isosceles, and very close to being an Equilateral triangle.

Now we need to draw Bindu which is a circle that **inscribes** Triangle T5d. Such a circle is found by bisecting each **angle** of the triangle, and where these bisectors meet that is the center of the circle to inscribe.

46. Draw the angle bisectors for T5d, and thus getting the center for new circle, draw the Bindu.
      46a. Using hatch, we fill the Bindu with Red color.
47 Cardinal Directions, switch off Cross Hair. We obtain the Sri Yantra core.

References

https://www.ashtangayoga.info/philosophy/sanskrit-and-devanagari/transliteration-tool/
https://www.learnsanskrit.cc/

Constructing the perpendicular bisector of a line segment
https://www.youtube.com/watch?v=3v5HL5nmrJE

Draw the Incenter and the Inscribed Circle of a triangle
https://www.youtube.com/watch?v=l2LOAXYU3UU

Draw the Circumcenter and the Circumcircle of a triangle
https://www.youtube.com/watch?v=kPsQvE7ZmS8

Draw the Sri Yantra (Patrick Flanagan)
https://sriyantraresearch.com/Construction/Flanagan/patrick_flanagan_method.htm

Draw the Sri Yantra by George Leoniak
https://knewgeometry.space/   https://www.youtube.com/watch?v=wqvYPCBEWWs

Draw the Sri Yantra by Zak Korvin
https://circle.gtryp.com/circle/tutorial-sri-yantra-version-1/   https://zkorvin.com/

Antonio Alessi https://pi-day.eye-of-revelation.org/Square-From-Circle.html#ratioceleste

S. K. Ramachandra Rao – The Tantra of Sri Chakra (Bhavanopanishat) - 1st – 1983 –Sharada Prakashana, Bangalore.

Sri Karapatra Swami, Sri Sitarama Kaviraja – श्रीविद्यारत्नाकरः - 8th – 2012 – Srividya Sadhana Pitha, Varanasi.

Ashwini Kumar Aggarwal – Sri Yantra with Golden Ratio Triangle and Inscriptions – 1st – 2023 – Devotees of Sri Sri Ravi Shankar Ashram, Punjab.

Epilogue

When one stops still and listens, a sound is heard. Felt within.

Perhaps it is that Om, the un-struck without origin.

सर्वे भवन्तु सुखिनः । सर्वे सन्तु निरामयाः ।

सर्वे भद्राणि पश्यन्तु । मा कश्चिद् दुःख भाग् भवेत् ॥

ॐ शान्तिः शान्तिः शान्तिः ॥

When faith has blossomed in life, Every step is led by the Divine.

Sri Sri Ravi Shankar

Om Namah Shivaya

जय गुरुदेव

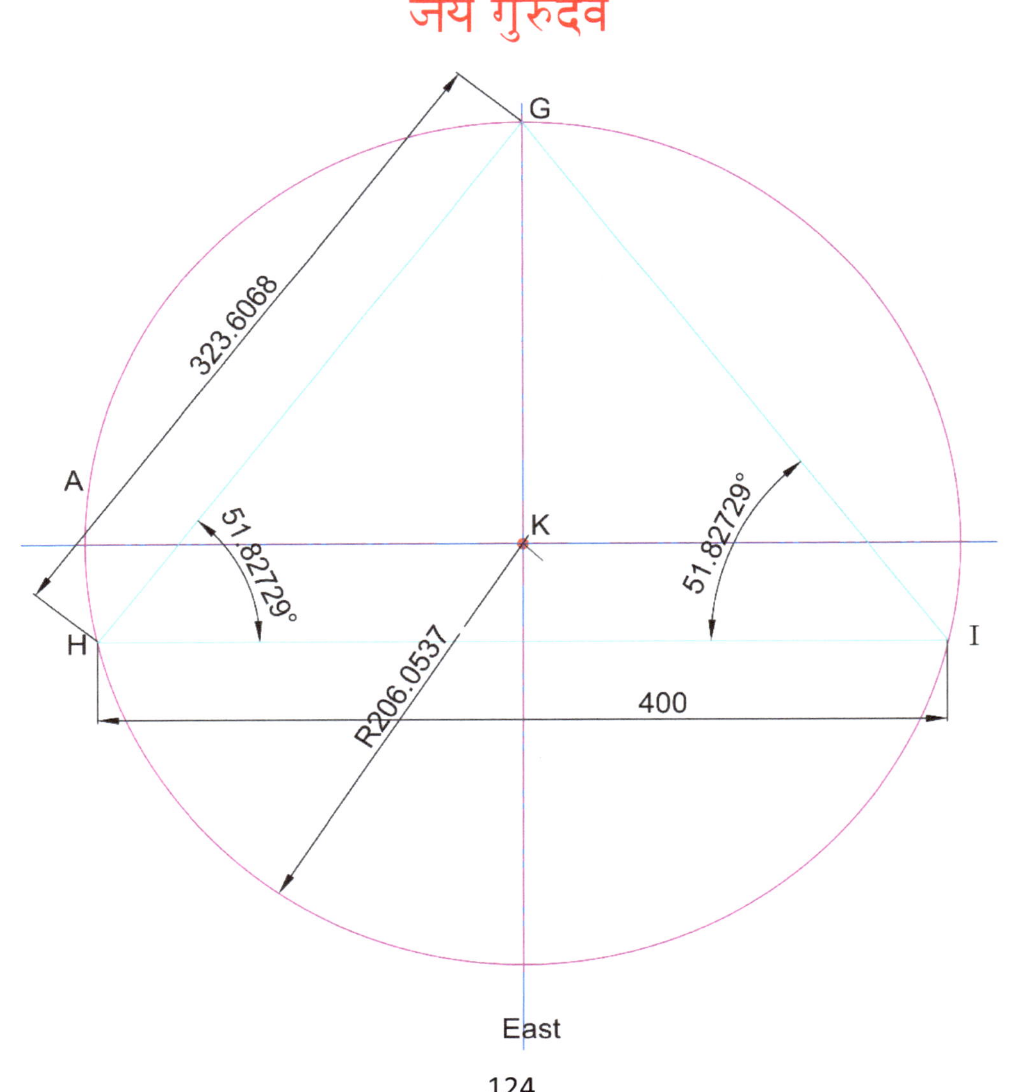

124

www.ingramcontent.com/pod-product-compliance
Lightning Source LLC
LaVergne TN
LVHW070046220726
843527LV00030B/482

9789395766623